Molecular Sensors for Cardiovascular Homeostasis

Molecular Sensors for Cardiovascular Homeostasis

Edited by

Donna H. Wang

Department of Medicine, Neuroscience, and Cell
and Molecular Biology Program
Michigan State University
East Lansing, Michigan, USA

Springer

Donna H. Wang
Department of Medicine
Michigan State University
East Lansing, MI 48824-1313
USA
donna.wang@ht.msu.edu

Cover illustration: Activation of TRPV1 by mechanical and chemical stimuli results in the release of CGRP and SP, which promote natriuresis and diuresis through their actions on the kidney. TRPV1 also affects kidney function via descending pathways from the CNS.

ISBN 1-4899-9199-9 ISBN 0-387-47530-3 (eBook)
ISBN 978-1-4899-9199-7 ISBN 978-0-387-47530-1 (eBook)

Printed on acid-free paper.

9 8 7 6 5 4 3 2 1

springer.com

Contents

Contributors

François M. Abboud, The Cardiovascular Research Center and the Departments of Internal Medicine and Molecular Physiology and Biophysics, Carver College of Medicine, University of Iowa, Iowa City, IA 52242, USA

Michael C. Andresen, Department of Physiology and Pharmacology, Oregon Health and Science University, Portland, OR 97239-3098, USA

Timothy W. Bailey, Department of Physiology and Pharmacology, Oregon Health and Science University, Portland, OR 97239-3098, USA

Abdul I. Barakat, Department of Mechanical and Aeronautical Engineering, University of California, Davis, CA 95616, USA

Dafni Bazopoulou, Institute of Molecular Biology and Biotechnology, Foundation for Research and Technology, Vassilika Vouton, Heraklion 71110, Crete, Greece

Christopher J. Benson, Department of Internal Medicine, Carver College of Medicine, University of Iowa, Iowa City, IA 52242, USA

Mark W. Chapleau, The Cardiovascular Research Center and the Departments of Internal Medicine and Molecular Physiology and Biophysics, Carver College of Medicine, University of Iowa, Iowa City, IA 52242; and the Veterans Affairs Medical Center, Iowa City, IA 52246, USA

David M. Cohen, Division of Nephrology and Hypertension, Oregon Health and Science University, and the Portland Veterans Affairs Medical Center, Portland, OR 97239, USA

Mark W. Doyle, Department of Biology, George Fox University, Newberg, OR 97132-2697, USA

Hans-Joachim Galla, Institut für Biochemie, Westfälische Wilhelms-Universität Münster, D-48149 Münster, Germany

Andrea Gojova, Department of Mechanical and Aeronautical Engineering, University of California, Davis, CA 95616, USA

Andreas Hinz, Institut für Biochemie, Westfälische Wilhelms-Universität Münster, D-48149 Münster, Germany

Young-Ho Jin, Department of Physiology and Pharmacology, Oregon Health and Science University, Portland, OR 97239-3098, USA

Deborah K. Lieu, Department of Mechanical and Aeronautical Engineering, University of California, Davis, CA 95616, USA

Yongjun Lu, The Cardiovascular Research Center and the Department of Internal Medicine, Carver College of Medicine, University of Iowa, Iowa City, IA 52242, USA

Edwin W. McCleskey, Vollum Institute, Oregon Health and Science University, Portland, OR 97239, USA

Tomoko Moriyama, Section of Cell Signaling, Okazaki Institute for Integrative Bioscience, National Institutes of Natural Sciences, Okazaki 444-8787, Japan

Andrea Olschewski, Medical University Graz, Department of Anesthesiology and Intensive Care Medicine, Auen Bruggerplatz 29, A-8036 Graz, Austria

Roger G. O'Neil, Department of Integrative Biology and Pharmacology, The University of Texas Health Science Center at Houston, Houston, TX 77030, USA

Jeffrey R. Sachs, B 316 Clinical Center, Department of Medicine, Michigan State University, East Lansing, MI 48824, USA

Nektarios Tavernarakis, Institute of Molecular Biology and Biotechnology, Foundation for Research and Technology, Vassilika Vouton, Heraklion 71110, Crete, Greece

Makoto Tominaga, Section of Cell Signaling, Okazaki Institute for Integrative Bioscience, National Institutes of Natural Sciences, Okazaki 444-8787, Japan

Giannis Voglis, Institute of Molecular Biology and Biotechnology, Foundation for Research and Technology, Vassilika Vouton, Heraklion 71110, Crete, Greece

Donna H. Wang, B 316 Clinical Center, Department of Medicine, Michigan State University, East Lansing, MI 48824-1313, USA

E. Kenneth Weir, Department of Medicine, VA Medical Center, Minneapolis, MN 55417, USA

Part I
The DEG/ENaC Family

1
The Role of DEG/ENaC Ion Channels in Sensory Mechanotransduction

Dafni Bazopoulou*, Giannis Voglis, and Nektarios Tavernarakis†

Abstract: All living organisms have the capacity to sense and respond to mechanical stimuli permeating their environment. Mechanosensory signaling constitutes the basis for the senses of touch and hearing and contributes fundamentally to development and homeostasis. Intense genetic, molecular, and elecrophysiological studies in organisms ranging from nematodes to mammals have highlighted members of the DEG/ENaC family of ion channels as strong candidates for the elusive metazoan mechanotransducer. These channels have also been implicated in several important processes including pain sensation, gametogenesis, sodium re-absorption, blood pressure regulation, and learning and memory. In this chapter, we review the evidence linking DEG/ENaC ion channels to mechanotransduction and discuss the emerging conceptual framework for a metazoan mechanosensory apparatus.

1.1. Introduction

Highly specialized macromolecular structures allow organisms to sense mechanical forces originating either from the surrounding environment or from within the organism itself. Such structures function as mechanotransducers, converting mechanical energy to biological signals. At the single-cell level, mechanical signaling underlies cell volume control and specialized responses such as the prevention of polyspermy in fertilization. At the level of the whole organism, mechanotransduction underlies processes as diverse as stretch-activated reflexes in vascular epithelium and smooth muscle, gravitaxis and turgor control in plants, tissue development and morphogenesis, and the senses of touch, hearing, and balance.

*Institute of Molecular Biology and Biotechnology, Foundation for Research and Technology, Heraklion 71110, Crete, GREECE
†Corresponding author: Institute of Molecular Biology and Biotechnology, Foundation for Research and Technology, Vassilika Vouton, P.O.Box 1527, Heraklion 71110, Crete, GREECE; tavernarakis@imbb.forth.gr

Elegant electrophysiological studies in several systems have established that mechanically-gated ion channels are the mediators of the response. For years, however, these channels have eluded intense cloning efforts. Why are these channels so particularly resistant to our exploitation? These channels are rare. In skin pads, mechanoreceptors are spread out so there are only 17,000 in the finger and palm skin pad.[1] This is an extremely low concentration. In the specialized hair cells of our ears, only a few hundred mechanically gated channels may exist. To make our prospects of directly encountering them even more slim, mechanosensory channels are embedded and intertwined with materials that attach them to the surrounding environment—contacts probably critical to function that are hard or even impossible to reconstitute or mimic in a heterologous system such as *Xenopus* oocytes, for example. Finally, there are no known biochemical reagents that interact with the mechanically gated channels with high specificity and high affinity, thwarting efforts for biochemical purification. Biochemical purification and structural analysis of an *E. coli* mechanosensitve channel, *Msc*L, has been accomplished,[2,3] but until recently, eukaryotic mechanosensitive ion channels have eluded cloning efforts, and thus little is understood of their structures and functions.

An alternative approach toward identifying the molecules that are involved in mechanotransduction is to identify them genetically. This approach has been particularly fruitful in the simple nematode, *Caenorhabditis elegans*.[4] Genetic dissection of touch transduction in this worm has led to the identification of several molecules that are likely to assemble into a mechanotransducing complex. These genetic studies revealed several genes that encode subunits of candidate mechanically gated ion channels involved in mediating touch transduction, proprioception, and coordinated locomotion.[5–8] These channel subunits belong to a large family of related proteins in *C. elegans* referred to as degenerins, because unusual gain-of-function mutations in several family members induce swelling or cell death.[9] *C. elegans* degenerins exhibit approximately 25–30% sequence identity to subunits of the vertebrate amiloride-sensitive epithelial Na^+ channels (ENaC), which are required for ion transport across epithelia, and acid-sensing ion channels that may contribute to pain perception and mechanosensation (ASICs, BNC).[10–13] Together, the *C. elegans* and vertebrate proteins define the DEG/ENaC (degenerin/epithelial sodium channel) family of ion channels.[11] Additional members of this large group of proteins are the snail FMRF-amide gated channel FaNaC,[14] the *Drosophila* ripped pocket and pickpocket (RPK and PPK)[15,16] and *C. elegans flr-1*.[17]

To summarize, members of the DEG/ENaC family have now been identified in organisms ranging from nematodes, snails, flies, and many vertebrates including humans, and are expressed in tissues as diverse as kidney and lung epithelia, muscle, and neurons. Intense genetic, molecular, and elecrophysiological studies have implicated these channels in mechanotransduction in nematodes, flies, and mammals.[11,18] Therefore, these proteins are strong candidates for a metazoan mechanosensitive ion channel (Table 1.1).

Here, we review the studies that led to the identification of nematode degenerins and discuss their role in mediating mechanosensitive behaviors in the worm. Furthermore, we correlate the mechanotransducer model that has emerged from

TABLE 1.1. DEG/ENaC proteins implicated in mechanotransduction

Protein	Expression pattern	Postulated function	Organism	Reference
DEL-1	Motorneurons Sensory neurons	Stretch sensitivity Proprioception	*Caenorhabditis elegans*	8
DEG-1	Interneurons Sensory neurons Muscle Hypodermis	Harsh touch sensitivity?	*Caenorhabditis elegans*	9
MEC-4	Touch receptor neurons	Touch sensitivity	*Caenorhabditis elegans*	5
MEC-10	Touch receptor neurons Other sensory neurons	Touch sensitivity	*Caenorhabditis elegans*	6
UNC-8	Motorneurons Interneurons Sensory neurons	Stretch sensitivity Proprioception	*Caenorhabditis elegans*	8
UNC-105	Muscle	Stretch sensitivity	*Caenorhabditis elegans*	7
PPK	Sensory dendrites of peripheral neurons	Touch sensitivity Proprioception	*Drosophila melanogaster*	15
DmNaCh	Multiple dendritic sensory neurons	Stretch sensitivity	*Drosophila melanogaster*	16
BNC1	Lanceolate nerve endings that surround the hair follicle	Touch sensitivity	*Mus musculus*	12
γENaC	Baroreceptor nerve terminals innervating the aortic arch and carotid sinus	Pressure sensitivity	*Rattus norvegicus*	19
ASIC3/ DRASIC	Dorsal root ganglia neurons; large-diameter mechanoreceptors; small-diameter peptidergic nociceptors	Mechanosensation; acid-evoked nociception	*Mus musculus*	20

investigations in *C. elegans* with recent findings in mammals, also implicating members of the DEG/ENaC family of ion channels in mechanotransduction. The totality of the evidence in such diverse species suggests that structurally related ion channels shape the core of a metazoan mechanotransducer.

1.2. Mechanosensory Signaling in *C. elegans*

C. elegans is a small (1 mm) soil-dwelling hermaphroditic nematode that completes a life cycle in 2.5 days at 25°C. Animals progress from a fertilized embryo through four larval stages to become egg-laying adults, and live for about 2 weeks. The simple body plan and transparent nature of both the egg and the cuticle of this nematode have facilitated exceptionally detailed developmental characterization of the animal. The complete sequence of cell divisions and the normal pattern of programmed cell deaths that occur as the fertilized egg develops into the 959-celled adult are both known.[21,22]

The anatomical characterization and understanding of neuronal connectivity in *C. elegans* are unparalleled in the metazoan world. Serial section electron microscopy has identified the pattern of synaptic connections made by each of the 302 neurons of the animal (including 5000 chemical synapses, 600 gap junctions, and 2000 neuromuscular junctions), so that the full "wiring diagram" of the animal is known.[23,24] Although the overall number of neurons is small, 118 different neuronal classes, including many neuronal types present in mammals, can be distinguished. Other animal model systems contain many more neurons of each class (there are about 10,000 more neurons in *Drosophila* with approximately the same repertoire of neuronal types). Overall, the broad range of genetic and molecular techniques applicable in the *C. elegans* model system allow a unique line of investigation into fundamental problems in biology such as mechanical signaling.

In the laboratory, *C. elegans* moves through a bacterial lawn on a petri plate with a readily observed sinusoidal motion. Interactions between excitatory and inhibitory motorneurons produce a pattern of alternating dorsal and ventral contractions.[25,26] Distinct classes of motorneurons control dorsal and ventral body muscles. To generate the sinusoidal pattern of movement, the contraction of the dorsal and ventral body muscles must be out of phase. For example, to turn the body dorsally, the dorsal muscles contract, while the opposing ventral muscles relax. The adult motor system involves five major types of ventral nerve cord motorneurons, defined by axon morphologies and patterns of synaptic connectivity. A motorneurons (12 VA and 9 DA), B motorneurons (11 VB and 7DB), D motorneurons (13 VD, 6 DD), AS motorneurons and VC motorneurons command body wall muscles arranged in four quadrants along the body axis.[25–27] Relatively little is known about how the sinusoidal wave is propagated along the body axis. Adjacent muscle cells are electrically coupled via gap junctions, which could couple excitation of adjacent body muscles. Alternatively, ventral cord motorneurons could promote wave propagation because gap junctions connect adjacent motorneurons of a given class.[23,24,28] A third possibility is that motorneurons could themselves act as stretch receptors so that contraction of body muscles could regulate adjacent motorneuron activities, thereby propagating the wave.[4,8]

When gently touched with an eyelash hair (typically attached to a toothpick) on the posterior, an animal will move forward; when touched on the anterior body, it will move backward. This gentle body touch is sensed by the six touch receptor neurons ALML/R (anterior lateral microtubule cell left, right), AVM (anterior ventral microtubule cell), and PLML/R (posterior lateral microtubule cell left, right; Fig. 1.1).

PVM (posterior ventral microtubule) is a neuron that is morphologically similar to the touch receptor neurons and expresses genes specific for touch receptor neurons but has been shown to be incapable of mediating a normal touch response by itself.[29–31] The touch receptors are situated so that their processes run longitudinally along the body wall embedded in the hypodermis adjacent to the cuticle. The position of the processes along the body axis correlates with the sensory field of the touch cell. Laser ablation of AVM and the ALMs, which have sensory receptor processes in the anterior half of the body, eliminates anterior touch sensitivity and laser ablation of the PLMs, which have posterior dendritic processes, eliminates

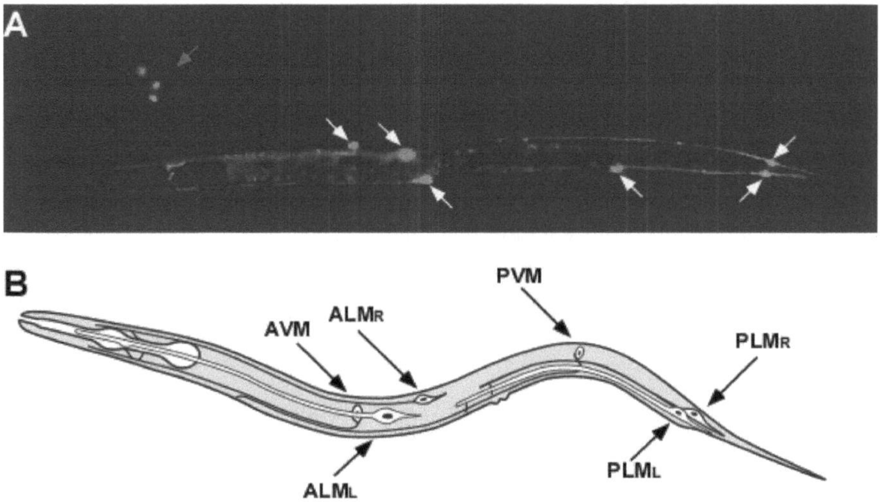

FIGURE 1.1. The *C. elegans* touch receptor neurons. (A) Visualization of touch receptors. Worms are expressing the green fluorescent protein (GFP) under the control of the *mec-4* promoter, which is active only in the six touch receptor neurons. Arrows indicate touch receptor cell bodies. Some touch receptor axons are apparent. (B) Schematic diagram, showing the position of the six touch receptor neurons in the body of the adult nematode. Note the two fields of touch sensitivity defined by the arrangement of these neurons along the body axis. The ALMs and AVM mediate the response to touch over the anterior field whereas PLMs mediate the response to touch over the posterior field. (See Color Plate 1 in Color Section)

posterior touch sensitivity. In addition to mediating touch avoidance, the touch receptor neurons appear to control the spontaneous rate of locomotion because animals that lack functional touch cells are lethargic. The mechanical stimuli that drive spontaneous locomotion are unknown, but could include encounters with objects in their environment or body stretch induced by locomotion itself. Touch receptor neurons have two distinguishing features. First, they are surrounded by a specialized extracellular matrix called the mantle which appears to attach the cell to the cuticle. Second, they are filled with unusual 15-protofilament microtubules.[32] Genetic studies suggest that both features are critical for the function of these neurons as receptors of body touch (reviewed in Ref. 4).

C. elegans displays several additional behaviors that are based on sensory mechanotransduction which have been characterized to a lesser extent. The nose of *C. elegans* is highly sensitive to mechanical stimuli. This region of the body is innervated by many sensory neurons that mediate mechanosensitivity. Responses to touch in the nose can be classified into two categories: the head-on collision response and the foraging and head withdrawal response.[33–36] Other mechanosensitive behaviors include the response to harsh mechanical stimuli, and the tap withdrawal reflex, where animals retreat in response to a tap on the culture plate.[37,38] Furthermore, mechanotransduction appears to also play a regulatory role in processes such as mating, egg laying, feeding, defecation, and maintenance of the pseudocoelomic body cavity pressure.[4,33] These behaviors add to the large repertoire of mechanosensitive phenomena, amenable to genetic and molecular dissection in the nematode.

1.2.1. Degenerins and Mechanotransduction in C. elegans

With the sequencing of the *C. elegans* genome now complete, it is possible to survey the entire gene family within this organism. Presently, 30 genes encoding members of the DEG/ENaC family have been identified in the *C. elegans* genome, seven of which have been genetically and molecularly characterized (*deg-1*, *del-1*, *flr-1*, *mec-4*, *mec-10*, *unc-8* and *unc-105*; Table 1.2).

TABLE 1.2. The current list of *C. elegans* DEG/ENaC family members and their chromosomal distribution. Genes have been listed alphabetically with the seven genetically characterized ones on top. Phenotypes are those of loss-of-function alleles. All 23 uncharacterized putative degenerin genes encode proteins with the sequence signature of amiloride-sensitive channels. However, some lack certain domains of typical DEG/ENaC ion channels (ND: Not Determined)

Gene name	ORF	Chromosome	Behavior/Phenotype	Reference
deg-1	C47C12.6	X	Touch abnormality	9
del-1	E02H4.1	X	Locomotory defects	8
mec-4	T01C8.7	X	Touch insensitivity	5
mec-10	F16F9.5	X	Touch insensitivity	6
flr-1	F02D10.5	X	Fluoride resistance	39
unc-8	R13A1.4	IV	Locomotory defects	8
unc-105	C41C4.5	II	Muscle function defects?	7
	C11E4.3	V		
	C11E4.4	X		
	C18B2.6	X		
	C24G7.1	I		
	C24G7.2	I		
	C24G7.4	I		
	C27C12.5	X		
	C46A5.2	X		
	F23B2.3	IV		
	F25D1.4	V		
	F26A3.6	I	ND	40
	F28A12.1	V		
	F55G1.12	IV		
	F59F3.4	IV		
	T21C9.3	V		
	T28B8.5	I		
	T28D9.7	II		
	T28F2.7	I		
	T28F4.2	I		
	Y69H2.2	V		
	Y69H2.11	V		
	Y69H2.13	V		
	ZK770.1	I		

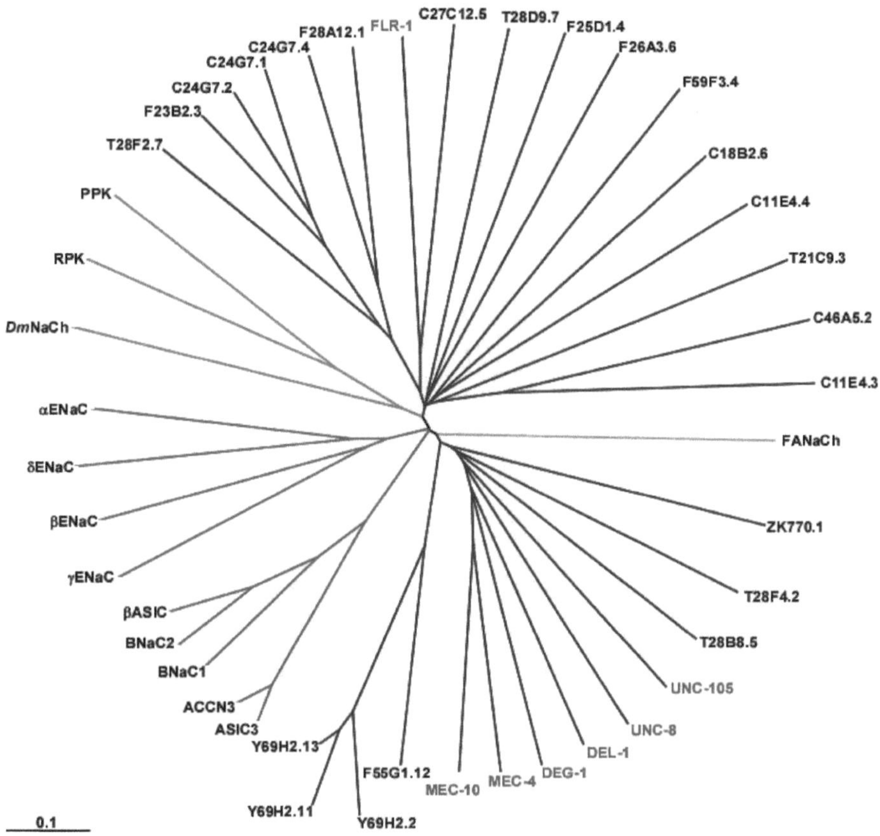

FIGURE 1.2. Phylogenetic relations among DEG/ENaC proteins in nematode degenerins are shown with blue lines. The current degenerin content of the complete nematode genome is included. The seven genetically characterized (DEG-1, DEL-1, FLR-1, MEC-4, MEC-10, UNC-8 and UNC-105) are shown in red. Representative DEG/ENaC proteins from a variety of organisms, ranging from snails to humans, are also included (mammalian: red lines; fly: green lines; snail: orange line). The scale bar denotes evolutionary distance equal to 0.1 nucleotide substitutions per site. (See Color Plate 2 in Color Section)

While DEG/ENaC proteins are involved in many diverse biological functions in different organisms, they share a highly conserved overall structure.[4,11,41] This strong conservation across species suggests that DEG/ENaC family members shared a common ancestor relatively early in evolution (Fig. 1.2).

The basic subunit structure may have been adapted to fit a range of biological needs by the addition or modification of functional domains. This conjecture can be tested by identifying and isolating such structural modules within DEG/ENaC ion channels.

DEG/ENaC proteins range from about 550 to 950 amino acids in length and share several distinguishing blocks of sequence similarity (Fig. 1.3). Subunit topology is

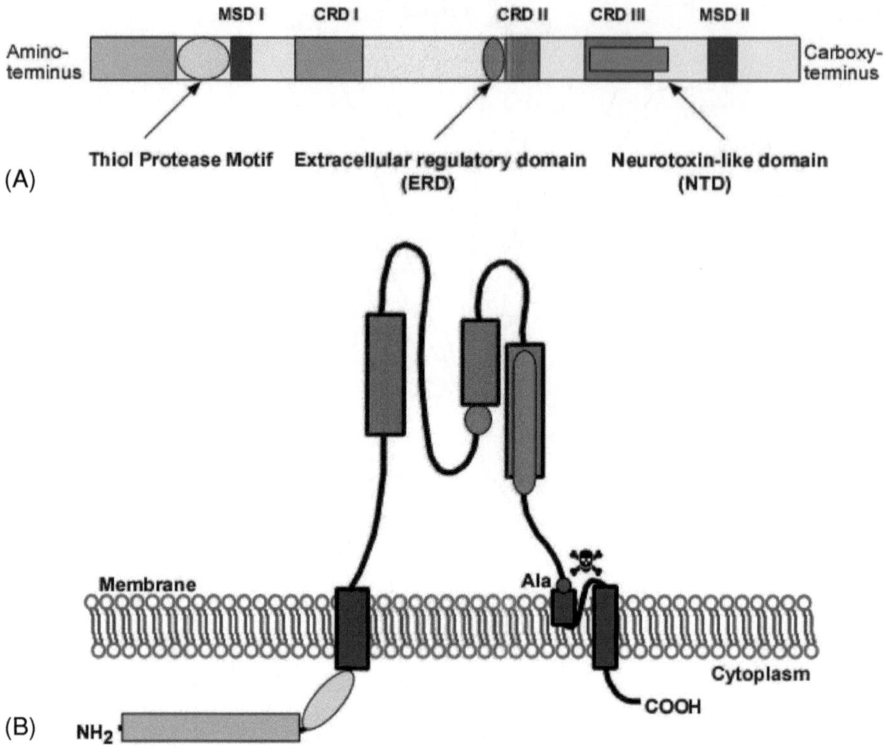

FIGURE 1.3. Schematic representation of DEG/ENaC ion channel subunit structure and topology. (A) Functional/structural domains. Colored boxes indicate defined channel modules. These include the two membrane-spanning domains (MSDs; dark-blue shading), and the three cysteine-rich domains (CRDs; red shading; the first CRD is absent in mammalian channels and is depicted by light red shading). The small light-blue oval depicts the putative extracellular regulatory domain (ERD). The green box overlapping with CRDIII denotes the neurotoxin-related domain (NTD). The conserved intracellular region with similarity to thiol-protease histidine active sites is shown in yellow. Shown in pink is the amino-terminal domain modeled based on protease pro-domains (see Fig. 1.7). (B) Transmembrane topology. Both termini are intracellular with the largest part of the protein situated outside the cell. The dot near MSDII represents the amino-acid position (Alanine 713 in MEC-4) affected in dominant, toxic degenerin mutants. (See Color Plate 3 in Color Section)

invariable: all DEG/ENaC family members have two membrane-spanning domains with cysteine-rich domains (CRDs, the most conserved is designated CRD3) situated between these two transmembrane segments.[18,42] DEG/ENaCs are situated in the membrane such that amino- and carboxy-termini project into the intracellular cytoplasm while most of the protein, including the CRDs, is extracellular (Fig. 1.3).[4,43] Highly conserved regions include the two membrane-spanning domains (MSD I and II), a short amino acid stretch before the first membrane-spanning domain, extracellular cysteine-rich domains (CRDs), an extracellular regulatory domain and a neurotoxin-related domain (NTD) before predicted transmembrane domain II.[42] The high degree of conservation of cysteine residues in these extracellular domains suggests that the tertiary structure of this region is critical to the function of most channel subunits and may mediate interactions with extracellular structures. Interestingly, the NTD is also distantly related to domains in several other proteins including the *Drosophila crumbs* protein, required for epithelial organization,[44] *agrin*, a basal lamina protein that mediates aggregation of acetylcholine channels,[45] and the selectins that participate in cell adhesion (such as ELAM-1).[46] The presence of related domains in proteins such as *crumbs* and *agrin* implies that such domains might act as interaction modules that mediate analogous interactions needed for tissue organization or protein clustering. We hypothesize that the appearance of neurotoxin-related domains in a specific class of ion channels may be the result of convergent evolution, driven by the requirement for high affinity interaction modules in these proteins.

Amino and carboxy termini are intracellular and a single large domain is positioned outside the cell (Fig. 1.3, Refs. 11, 47). The more amino-terminal of the two membrane-spanning domains (MSDI) is generally hydrophobic, whereas the more carboxy-terminal of these (MSDII) is amphipathic.[48,49] In general, MSDI is not distinguished by any striking sequence feature except for the strict conservation of a tryptophan residue (corresponding to position W111 in MEC-4), and the strong conservation of a Gln/Asn residue (corresponding to position N125 in MEC-4). MSDII is more distinctive, exhibiting strong conservation of hydrophilic residues (consensus GLWxGxSxxTxxE) that has been implicated in pore function.[48] The short highly conserved region before the minimal transmembrane domain is thought to loop back into the membrane to contribute to the channel pore.[41,50,51] The extended MSDII homology region (loop + transmembrane part) can be considered a defining characteristic of DEG/ENaC family members.

Below we discuss two nematode mechanosensitive behaviors that involve degenerins: the gentle body-touch response and locomotion. Furthermore, we highlight similarities in the structure and function of these proteins.

1.2.1.1. The Gentle Touch Response

Approximately 15 genes have been identified by genetic analysis, which, when mutated, specifically disrupt gentle body touch sensation. These genes are therefore thought to encode candidate mediators of touch sensitivity (these genes were named *mec* genes because when they are defective, animals are *mech*anosensory abnormal).[52] Almost all of the *mec* genes have now been molecularly identified, and most of them encode proteins postulated to make up a touch-transducing complex.[53,54] The core elements of this mechanosensory complex are the channel subunits MEC-4 and MEC-10, which can interact genetically and physically.[55,56] Both these proteins are DEG/ENaC family members.

MEC-4, MEC-10 and several related nematode degenerins have a second, unusual property: specific amino acid substitutions in these proteins result in aberrant channels that induce the swelling and subsequent necrotic death of the cells in which they are expressed.[57] This pathological property is the reason that proteins of this subfamily were originally called degenerins.[9] For example, unusual gain-of-function (dominant; *d*) mutations in the *mec-4* gene induce degeneration of the six touch receptor neurons required for the sensation of gentle touch to the body. In contrast, most *mec-4* mutations are recessive loss-of-function mutations that disrupt body touch sensitivity without affecting touch receptor ultrastructure or viability (reviewed in Ref. 4).

Evidence that MEC-4 and MEC-10 co-assemble into the same channel complex include the following: (1) MEC-4 and MEC-10 subunits are co-expressed in the touch receptor neurons;[6] (2) MEC-4 and MEC-10 proteins translated *in vitro* in the presence of microsomes can co-immunoprecipitate;[56] and (3) genetic interactions between *mec-4* and *mec-10* have been observed.[53] For example, *mec-10* can be engineered to encode a death-inducing amino acid substitution *mec-10* (A673V).[6] However, if *mec-10* (A673V) is introduced into a *mec-4* loss-of-function background, neurodegeneration does not occur. This result is consistent with the hypothesis that MEC-10 cannot form a functional channel in the absence of MEC-4. Genetic experiments also suggest that MEC-4 subunits interact with each other. The toxic protein MEC-4 (A713V) encoded by the *mec-4(d)* allele can kill cells even if it is co-expressed with wild-type MEC-4(+) (as occurs in a *trans* heterozygote of genotype *mec-4(d)/mec-4(+)*). However, if toxic MEC-4 (A713V) is co-expressed with a specific *mec-4* allele that encodes a single amino acid substitution in MSDII (e.g., *mec-4(d)/mec-4* (E732K)), neurodegeneration is partially suppressed.[53] Because one MEC-4 subunit can interfere with the activity of another, it can be inferred that there may be more than one MEC-4 subunit in the channel complex.

Amino acids on the polar face of amphipathic transmembrane MSDII are highly conserved and are essential for *mec-4* function.[48] Consistent with the idea that these residues project into the channel lumen to influence ion conductance, amino acid substitutions in the candidate pore domain (predicted to disrupt ion influx) block or delay degeneration when the channel-opening, Ala713Val substitution is also present in MEC-4.[11,48,51] Electrophysiological characterization of rat and rat/nematode chimeras supports the hypothesis that MSDII constitutes a pore-lining domain and that highly conserved hydrophilic residues in MSDII face into the channel lumen to influence ion flow.[58,59]

mec-4(d) alleles encode substitutions for a conserved alanine that is positioned extracellularly, adjacent to pore-lining membrane-spanning domain (Fig. 1.3; alanine 713 for MEC-4[5]). The size of the amino acid sidechain at this position is correlated with toxicity. Substitution of a small sidechain amino acid does not induce degeneration, whereas replacement of the Ala with a large sidechain amino acid is toxic. This suggests that steric hindrance plays a role in the degeneration mechanism and supports the following working model for *mec-4(d)*-induced degeneration: MEC-4 channels, like other channels, can assume alternative open and closed conformations. In adopting the closed conformation, the sidechain of the amino acid at MEC-4 position 713 is proposed to come into close proximity to another part of the channel. Steric interference conferred by a bulky amino acid sidechain prevents such an approach, causing the channel to close less effectively. Increased cation influx initiates neurodegeneration. That ion influx is critical for degeneration is supported by the fact that amino acid substitutions that disrupt the channel conducting pore can prevent neurodegeneration when present *in cis* to the A713 substitution. Other *C. elegans* family members (e.g., *deg-1* and *mec-10*) can be altered by analogous amino acid substitutions to induce neurodegeneration.[6,9] In addition, large sidechain substitutions at the analogous position in some neuronally expressed mammalian superfamily members do markedly increase channel conductance.[60,61]

Interestingly, the cell death that occurs appears to involve more than the burst of a cell in response to osmotic imbalance.[62] Rather, it appears that the necrotic cell death induced by these channels may activate a death program that is similar in several respects to that associated with the excitotoxic cell death that occurs in higher organisms in response to injury, in stroke, and so on. Electron microscopy studies of degenerating nematode neurons that express the toxic *mec-4(d)* allele have revealed a series of distinct events that take place during degeneration, involving extensive membrane endocytosis and degradation of cellular components.[63] Thus, the toxic degenerin mutations provide the means with which to examine the molecular genetics of injury-induced cell death in a highly manipulable experimental organism.

1.2.1.2. Sinusoidal Locomotion

Unusual, *s*emi-*d*ominant gain-of-function mutations in another degenerin gene, *unc-8*, (*unc-8(sd)*) induce transient neuronal swelling and severe lack of coordination.[64–66] *unc-8* encodes a degenerin expressed in several motor neuron classes and in some interneurons and nose touch sensory neurons.[8] Interestingly, semi-dominant *unc-8* alleles alter an amino acid in the region hypothesized to be an extracellular channel-closing domain defined in studies of *deg-1* and *mec-4* degenerins.[8,67] The genetics of *unc-8* are further similar to those of *mec-4* and *mec-10*; specific *unc-8* alleles can suppress or enhance *unc-8(sd)* mutations *in trans*, suggesting that UNC-8::UNC-8 interactions occur. Another degenerin family member, *del-1*(for *deg*enerin-*l*ike) is co-expressed in a subset of neurons that express *unc-8* (the VA and VB motor neurons) and is likely to assemble into a channel complex with UNC-8 in these cells.[8]

What function does the UNC-8 degenerin channel serve in motorneurons? *unc-8* null mutants have a subtle locomotion defect.[8] Wild-type animals move through an *E. coli* lawn with a characteristic sinusoidal pattern. *unc-8* null mutants inscribe a path in an *E. coli* lawn that is markedly reduced in both wavelength and amplitude as compared to wild-type (Fig. 1.4).

This phenotype indicates that the UNC-8 degenerin channel functions to modulate the locomotory trajectory of the animal.

How does the UNC-8 motor neuron channel influence locomotion? One highly interesting morphological feature of some motorneurons (in particular, the VA and VB motorneurons that co-express *unc-8* and *del-1*) is that their processes include

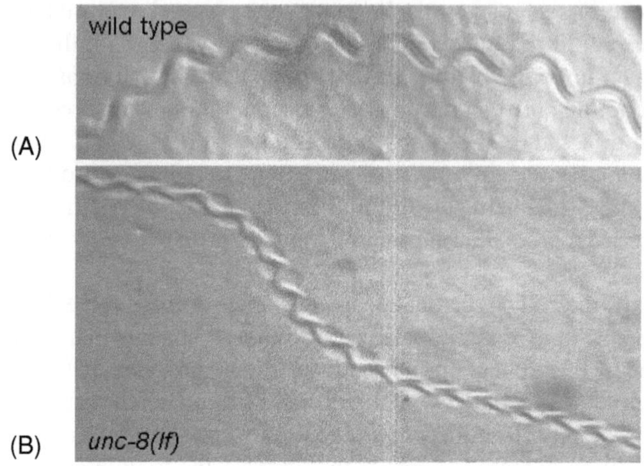

FIGURE 1.4. Proprioception in the nematode. (A) Wild-type animals inscribe a sinusoidal track as they move on an agar plate evenly covered with an *E. coli* bacterial lawn. (B) The characteristic properties (amplitude and wavelength) of tracks inscribed by *unc-8(lf)* mutants are drastically reduced. (See Color Plate 4 in Color Section)

extended regions that do not participate in neuromuscular junctions or neuronal synapses. These "undifferentiated" process regions have been hypothesized to be stretch-sensitive (discussed in Ref. 23). Given the morphological features of certain motor neurons and the sequence similarity of UNC-8 and DEL-1 to candidate mechanically gated channels, we have proposed that these subunits co-assemble into a stretch-sensitive channel that might be localized to the undifferentiated regions of the motor neuron process[8] reviewed in Ref. 4. When activated by the localized body stretch that occurs during locomotion, this motor neuron channel potentiates signaling at the neuromuscular junction, which is situated at a distance from the site of the stretch stimulus (Fig. 1.5).

The stretch signal enhances motorneuron excitation of muscle, increasing the strength and duration of the pending muscle contraction and directing a full size body turn. In the absence of the stretch activation, the body wave and locomotion still occur, but with significantly reduced amplitude because the potentiating stretch signal is not transmitted. This model bears similarity to the chain reflex mechanism of movement pattern generation. However, it does not exclude a central oscillator that would be responsible for the rhythmic locomotion. Instead, we suggest that the output of such an oscillator is further enhanced and modulated by stretch sensitive motorneurons.

One important corollary of the *unc-8* mutant studies is that the UNC-8 channel does not appear to be essential for motor neuron function; if this were the case, animals lacking the *unc-8* gene would be severely paralyzed. This observation strengthens the argument that degenerin channels function directly in mechanotransduction rather than merely serving to maintain the osmotic environment so that other channels can function. As is true for the MEC-4 and MEC-10 touch

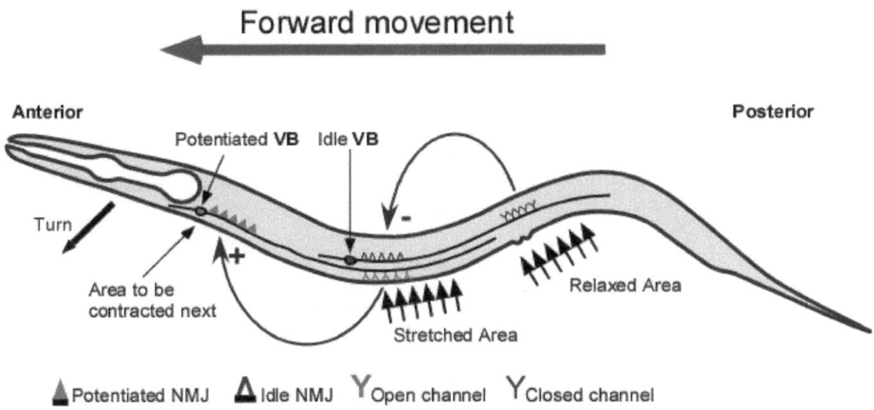

FIGURE 1.5. A model for UNC-8 involvement in stretch-regulated control of locomotion. Schematic diagram of potentiated and inactive VB class motor neurons. Neuro-muscular junctions (signified by triangles) are made near the cell body. Mechanically-activated channels postulated to include UNC-8 (and, possibly in VB motor neurons, DEL-1) subunits (signified by Y figures) are hypothesized to be concentrated at the synapse-free, undifferentiated ends of the VB neuron. Mechanically gated channels could potentiate local excitation of muscle. Body stretch is postulated to activate mechanically gated channels that potentiate the motor neuron signal that excites a specific muscle field. A strong muscle contraction results in a sustained body turn. In *unc-8(lf)* mutants, VB motor neurons lack the stretch-sensitive component that potentiates their signaling and consequently elicit a muscle contraction that is shortened in intensity or duration so that the body turns less deeply. Note that although we depict VB as an example of one motor neuron class that affects locomotion, other motor neuron classes must also be involved in the modification of locomotion in response to body stretch. Sequential activation of motor neurons that are distributed along the ventral nerve cord and signal nonoverlapping groups of muscles, amplifies and propagates the sinusoidal body wave (NMJ: neuromuscular junction). (See Color Plate 6 in Color Section)

receptor channel, the model of UNC-8 and DEL-1 function that is based on mutant phenotypes, cell morphologies and molecular properties of degenerins remains to be tested by determining subcellular channel localization, subunit associations and, most importantly, channel gating properties.

1.2.2. A Model for the Nematode Mechanotransducer

The features of cloned touch cell and motorneuron structural genes together with genetic molecular and electrophysiological data that suggest interactions between them constitute the basis of a model for the nematode mechanotransducing complex (Fig. 1.6).

The central component of the mechanotransduction apparatus is the putative mechanosensitive ion channel that includes multiple MEC-4 and MEC-10 subunits in the case of touch receptor neurons, and UNC-8 and DEL-1 subunits in the

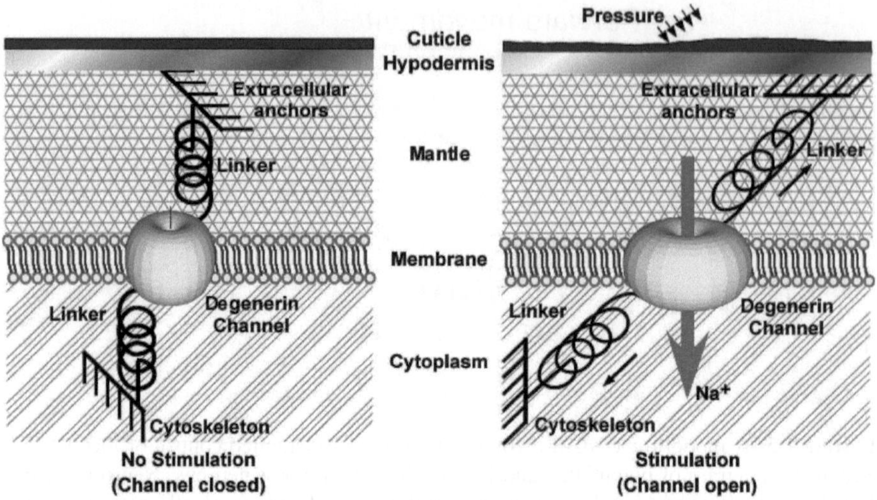

FIGURE 1.6. A mechanotransducing complex in *C. elegans* touch receptor neurons. In the absence of mechanical stimulation the channel is closed and therefore the sensory neuron is idle. Application of a mechanical force to the body of the animal results in distortion of a network of interacting molecules that opens the degenerin channel. Na^+ influx depolarizes the neuron initiating the perceptory integration of the stimulus. (See Color Plate 7 in Color Section)

case of motorneurons (reviewed in Refs. 4, 68). These subunits assemble to form a channel pore that is lined by the hydrophilic residues of membrane-spanning domain II. Subunits adopt a topology in which the cysteine-rich and neurotoxin-related domains extend into the specialized extracellular matrix outside the touch cell and the amino- and carboxy-termini project into the cytoplasm. Regulated gating depends on mechanical forces exerted on the channel. Tension is delivered by tethering the extracellular channel domains to the specialized extracellular matrix and anchoring intracellular domains to the microtubule cytoskeleton. Outside the cell, channel subunits may contact extracellular matrix components (such as *mec-1*, *mec-5* and/or *mec-9* in the case of the touch receptor mantle[55,67,69]). Inside the cell, channel subunits may interact with the cytoskeleton either directly or via protein links (such as MEC-2 in the touch receptor neurons or UNC-1 in motorneurons[56,70]).

Sequence analysis of recessive loss-of-function *mec-4* alleles has highlighted two regions of MEC-4, which appear especially important in channel gating. Amino acid substitutions that disrupt MEC-4 function cluster within a conserved region that is situated on the intracellular side, close to MSDI.[49] This region of the channel could interact with cytoskeletal proteins (Fig. 1.7).

Interestingly, the effects of semi-dominant alleles of *unc-8* can be completely blocked by mutations in this conserved region, highlighting its functional importance.[8,65,66] This suppression is observed both when such mutations reside

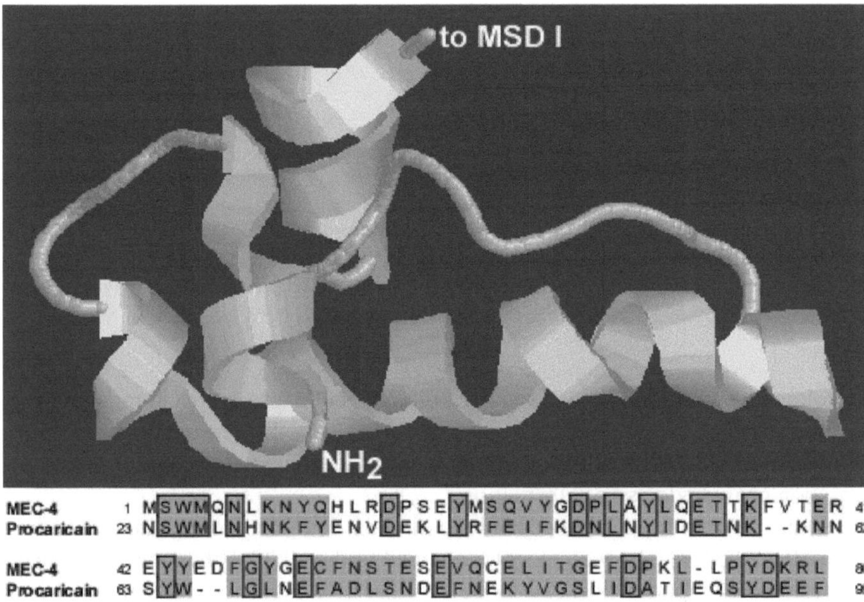

FIGURE 1.7. A three-dimensional model of the extreme, intracellular amino-terminus of MEC-4. The domain has been modeled by homology to the protease procaricain (the relevant alignment is shown at the bottom). The resulting structure appears to have the capacity for protein-protein interactions with a potential hydrophobic surface.[71] (See Color Plate 8 in Color Section)

in cis, on the same protein molecule as the semi-dominant mutations or *in trans*, on different co-expressed genes, as observed in heterozygote animals carrying a semi-dominant allele on one chromosome and a mutation in the conserved intracellular amino terminal region on the other.[65,71] Such a pattern of genetic suppression suggests that UNC-8 proteins interact to form a dimeric or multimeric complex where more than one molecules associate to form a channel. The conserved intracellular amino terminal region could play a role in facilitating such interactions. A second hot-spot for channel-inactivating substitutions is situated near and within NTD or within CRDII.[42] This is a candidate region for interaction of the channel with the extracellular matrix.

The mechanosensory apparatus encompassing MEC-4 and MEC-10 subunits appears to be localized at the long processes of touch receptor neurons (Fig. 1.8). A touch stimulus either could deform the microtubule network, or could perturb the mantle connections to deliver the gating stimulus (Fig. 1.6). In both scenarios, Na^+ influx would activate the touch receptor to signal the appropriate locomotory response. This is an attractive hypothesis, but confirmation has been stonewalled by the technical challenge of stimulating and recording directly from the *C. elegans* touch neurons, which are tiny (soma on the order of 1 μm) and embedded in the hypodermis. Furthermore, reconstitution of the mechanotransducing complex in a

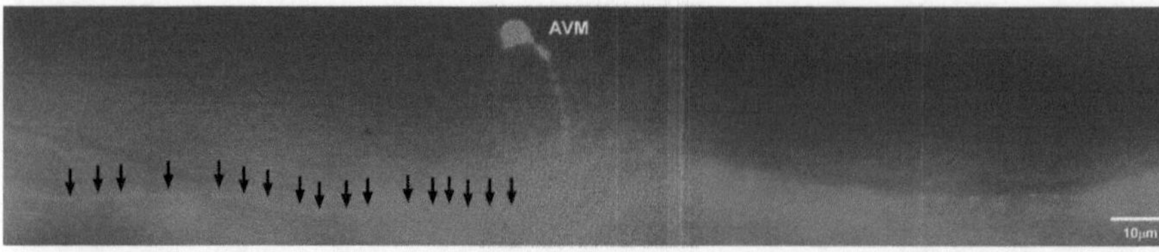

FIGURE 1.8. Punctate localization of a putative mechanosensitive ion channel subunit. Image of an AVM touch receptor neuron expressing a GFP-tagged MEC-4 protein. Fluorescence is unevenly distributed along the process of the neuron in distinct puncta, which may represent the location of the mechanotransducing apparatus. (See Color Plate 5 in Color Section)

heterologous system is likely to require both channel expression and regeneration of gating contacts, which would be no small feat. Nonetheless, ongoing efforts to surmount technical difficulties in direct recording from nematode sensory neurons may soon provide decisive information.

Because it has not yet been possible to directly demonstrate mechanical gating of the MEC-4/MEC-10 touch receptor channel or the UNC-8 channels using electrophysiological approaches, two models for the biological activities of degenerin channels have been considered.[4] In the simplest model, the degenerin channel mediates mechanotransduction directly. The alternative model is that the degenerin channel acts indirectly to maintain a required osmotic balance within a neuron so that a mechanosensitive channel, yet to be identified, can function. In the case of the touch receptor channel, the absence of either MEC-4 or MEC-10 renders the mechanosensory neuron nonfunctional, making it impossible to distinguish between the two alternative hypotheses. The situation with the UNC-8 channel is different. It is clear from the phenotype of *unc-8* null mutants that the majority of neurons that express *unc-8* must remain functional in the absence of *unc-8* activity.[8] Our understanding of neuronal circuitry and characterized behavioral mutants argues that if these neurons were not functional, *unc-8* null mutants would exhibit severely defective locomotion. Given that *unc-8* null mutants move in a manner only marginally different from wild-type animals, the case that the UNC-8 channel maintains an osmotic milieu required for the function of other neuronal channels is weakened. One caveat to this discussion is that we cannot rule out the possibility that a functionally redundant and as yet unidentified degenerin family member might be co-expressed with *unc-8* and could nearly compensate for its absence.

The model proposed for mechanotransduction in the touch receptor neurons and motorneurons of *C. elegans* shares the same underlying principle and features of the proposed gating mechanism of mechanosensory ion channels in *Drosophila* sensory bristles, and the channels that respond to auditory stimuli in the hair cells of the vertebrate inner ear.[72–75] Hair cells have bundles of a few hundred stereocilia on their apical surface, which mediate sensory transduction. Stereocilia are connected at their distal ends to neighboring stereocilia by filaments called tip links. The integrity of the tip links is essential for channel opening and the mechanosensitive channels appear to be situated at the ends of the stereocilia, near the connecting tip links. Directional deflection of the stereocilia relative to one another introduces tension on the tip links, which is proposed to open the mechanosensitive hair cell channels directly.

1.3. DEG/ENaC Mechanosensitive Channels in *Drosophila*

Two members of the DEG/ENaC family of ion channels have been implicated in mechanotransduction in *Drosophila*, pickpocket (PPK) and DmNaCh. PPK was found in the sensory dendrites of a subset of peripheral neurons in late-stage embryos and early larvae. In insects, such multiple dendritic neurons play key roles in touch sensation and proprioception and their morphology resembles human mechanosensory free nerve endings. These results suggest that PPK may be a channel subunit involved in mechanosensation.[15] DmNaCh is expressed in the dendritic arbor subtype of multiple dendritic (md) sensory neurons in the *Drosophila* peripheral nervous system. DmNaCh mRNA was first detected during late embryogenesis. While the origin and specification of md neurons are well documented, their roles are still poorly understood. They could function as stretch or touch receptors, raising the possibility that DmNaC could also be involved in mechanotransduction.[16]

1.4. DEG/ENaC Mechanosensitive Channels in Mammals

An increasing amount of evidence suggests that some mammalian DEG/ENaC proteins may play a role in mechanosensation similarly to their nematode counterparts. Mammalian members of the DEG/ENaC family fall into two classes. The first class includes α, β, and γ, ENaC, which form a multimeric epithelial channel that performs critical functions in Na^+ reabsorption in the kidney and in fluid clearance in neonatal lung.[11] A second ENaC subfamily includes proteins more prevalently found in neurons.[76,77] Channels in this group are known to be gated by protons and have thus been classified as the ASIC family (acid-sensing ion channel).[78,79] The acid-sensitive gating properties of certain neuronally expressed homomeric or heteromeric channels fuel speculation that these ASIC channels might respond to the local acidosis that occurs in injured or inflamed tissue and thus play critical roles in nociception.[80,81]

ENaC proteins are expressed in epithelial cells of the colon, in the apical membranes of specific epithelial structures such as the renal cortical collecting ducts (CCDs) and in airways cells. ENaC subunits are also detected in epithelial and nonepithelial structures in the rat cochlea.[11] ENaC proteins in endothelial cells are likely to be involved in responses to mechanical stimuli. Because of their location, endothelial cells are subjected to various types of mechanical forces such as osmotic pressure, hydrostatic pressure, fluid shear stress, mechanical stress by blood flow, air breathing and air vibration on the hair cells of the inner ear. Because ENaC channels are sensitive to membrane stretch and changes to the cell volume it is possible that ENaC conductance is regulated by these biomechanical forces.

There are strong indications that ENaC subunits may be components of the baroreceptor mechanotransducer, one of the most potent regulators of anterial pressure and neurohumoral control of the circulation.[19,82] Baroreceptors innervate the aortic arch and carotid sinuses and are activated by pressure-induced vessel-wall stretch. Furthermore, it has been shown that β- and γENaC, but not αENaC, are located in tactile sensory receptors in the hairless skin of the rat paw, suggesting that these subunits may be components of a mechanosensory receptor for touch.[83] ENaC immunoreactivity was also detected in mechanosensory lanceolate nerve endings of the rat mystacial pad in the vibrassae (whisker).

Members of the ASIC (acid-sensing ion channel) subgroup of the DEG/ENaC family have been implicated in mechanotransduction in mammals. BNC1 (brain Na^+ channel; also known as MDEG, BNaC1, ASIC2)[13,84–86] has emerged as promising candidate for a mechanosensitive channel; it is the ASIC member most similar in amino acid sequence to nematode MEC-10 and can be can be genetically altered analogously to MEC-4 and MEC-10 to generate hyperactive, toxic channels.[86,87] There exist two splice variants of BNC1 (a and b; also known as MDEG1 and MDEG2[79,85,88]) that share a common carboxy-terminal half.

In rodent hairy skin, several specialized nerve termini function as mechanoreceptors, including rapidly adapting (RA), slowly adapting (SA), and D-hair receptors.[1] Antisera raised against BNC1 identifies large numbers of central nervous system neurons, but also reveals that BNC1 specifically localizes to the palisades of lanceolate nerve terminals—fine parallel processes projected in the hair follicle and surrounding the hair shaft, a likely site for sensation of hair movement.[12] These nerve terminals house one type of rapidly adapting mechanoreceptor. Interestingly, in these studies BNC1 immunoreactivity is not prevalent in other nerve termini intimately associated with the hair follicle and implicated in mechanotransduction, such as the pilo-Ruffini endings that also circle the hair shaft terminal, or other mechanoreceptors or nociceptors.[12] The specific subcellular localization is striking in that many large- and small-diameter dorsal root ganglion neurons express messages homologous to BNC1, yet the protein is localized to only a few mechanosensory termini.[12] Broad transcript expression in large and small diameter neurons, but rare localization of the protein in nerve termini, has been observed for ENaCs in the dorsal root ganglion and in baroceptor neurons.[13,83,89,90] Such specificity indicates that mechanisms for localized or selective positioning of DEG/ENaC channels are operative in peripheral neurons. Alternatively, channel proteins may not sufficiently concentrated to be easily detected by immunological methods, a characteristic of typical mechanoreceptor channels.

Does BNC1 play a role in mechanosensation or nociception? Either (or both) is plausible because BNC1 is detectable in both large-diameter neurons (mostly mechanosensitive neurons) and small-diameter neurons (mostly nociceptors) of the dorsal root ganglion.[12,13] Generation of a BNC1 mouse knockout enabled testing of these possibilities. Both splice variants of BNC1 were eliminated in this mouse.[12] At a gross level, the BNC1 null mice appear generally normal in development, size, fertility, and behavior.

To address a potential function in mechanotransduction, detailed characterization of skin sensory neurons was performed on a skin-nerve preparation in which nerve terminals are tested for response to an applied displacement force.[12] This hairy skin preparation houses all five specialized mechanoreceptor types, classified based on their electrophysiological properties: rapidly adapting (RA) low-threshold mechanoreceptors, slowly adapting (SA) low-threshold mechanoreceptors, D-hair mechanoreceptors, A-fiber mechano-nociceptors, and polymodal C-fiber mechano-nociceptors.[1] In BNC1 −/− animals, neither the stimulus-response curves nor the median force required to activate D-hair mechanoreceptors, A-fiber mechano-nociceptors and C-fiber mechano-nociceptors is altered, compared to BNC1+/+ controls.[12]

Likewise, all efforts to test for changes in acid-induced responses and nociception in dorsal root ganglion neurons and poly-modal C fibers failed to indicate an essential role for BNC1 in modulating H^+-gated currents in these cells. In contrast, a striking change in the function of RA and SA low-threshold mechanoreceptors was observed in the BNC1 null mutant. Although the minimal force detectable for activation of these mechanoreceptors remains the same, the stimulus-response curve for RA, and to a lesser extent SA, BCN1–/– neurons is significantly different.[12]

In wild-type nerve terminals, increasing the force exerted on the fiber elicits increasing numbers of action potentials. Mutant neurons still respond to displacement, but produce fewer action potentials over a comparable range of stimuli. Interestingly, the effects on the action potential do not appear to result from developmental defects in the neurons involved. There are no apparent differences in the proportion of RA and SA fibers in skin preparations of wild type and mutant mice. In addition, the number and morphology of lanceolate fibers (one, but not the only, type of RA receptor) is similar in BNC1+/+ and –/– animals.[12] Similarly, in nematodes the touch receptor neurons can develop normally in the absence of MEC-4 channels. Also important is that the defects in action potential firing in the BNC1 mutant appear to affect something other than the capacity to generate an action potential. Injection currents required to elicit action potentials in cultured low-threshold mechanoreceptor neurons from BNC1+/+ and BNC1–/– mice are similar. Because the basic capacity to convert a depolarizing inward current to an action potential appears to be normal in the BNC1–/– sensory neurons, it appears that the problem in BNC1–/– neurons is the actual generation of a mechanically induced depolarizing potential, consistent with the hypothesis that BNC1 participates directly in a mechanosensitive channel.

The consequences of the BNC1 channel deficiency, although somewhat modest at first glance, may be of profound biological importance because in humans the dynamic sensitivity of RA and SA receptors is thought to be critical for perception and discrimination of touch sensation.[91,92] Why might the response be modified rather than eliminated in mechanosensitive neurons of the BNC1 knockout? One plausible reason is that DEG/ENaC channels are most often hetero-multimeric and

BNC1 might act more as a modulatory subunit than as the core of a mechanotransducing complex, much as β and γENaC are less critical than αENaC function in kidney epithelia. Alternatively, different DEG/ENaC channels (or other channels) may perform redundant functions in the same neurons. Consistent with this possibility, ENaC subunits have been immunologically detected in neurons expressing BNC1, suggesting ENaC subunits could be components of mechanotransducing channels in neurons as well.

In addition to BNC1, ASIC3 (also known as DRASIC) is a good candidate for transduction of mechanical stimuli because it is expressed in mechanosensors and nociceptors.[20] ASIC3 was detected in DRG neurons, both small- and large-diameter cells, in rapidly adapting (RA) mechanoreceptors (fibers of Meissner corpuscles, guard hair follicles), in slowly adapting mechanoreceptors (Merker cells) and in AM mechano-nociceptors. RA mechanoreceptors respond to light touch and AM nociceptors respond to noxious stimuli (pain sensation). Loss of ASIC3 in mice enhanced mechanosensitivity in RA mechanoreceptors and reduced mechanosensitivity in AM nociceptors. $ASIC3^{-/-}$ mice displayed decreased responses in acidic and noxious heat stimulus. Thus, it is suggested that ASIC3 participates in mechanosensation in response to mechanical stimuli and changes in pH.[20]

1.5. Outlook

Genetic analyses have been highly successful in identifying genes needed for mechanosensitive behaviors.[72,73,93–95] Still, limitations of the genetic approach to dissection of mechanotransduction mechanisms should be mentioned. Genes that encode products needed for the activities of mechanotransducing complexes in multiple cell types or that perform multiple cellular functions might have evaded genetic detection because mutations in such genes would be expected to be severely uncoordinated or even lethal. Indeed, many mutations that affect mechanosensation in *Drosophila* render animals severely uncoordinated and nearly inviable.[94,96] Moreover, genes whose functions are redundantly encoded cannot be readily identified in genetic screens. Thus, additional cellular proteins essential for the mechanotransducing complex in the well-studied *C. elegans* body touch receptor neurons may still remain to be identified.

The detailed model for mechanotransduction in *C. elegans* neurons, accommodates genetic data and molecular properties of cloned genes. This model remains to be tested by determining subcellular channel localization, subunit associations and, most importantly, channel-gating properties. The proposed direct interactions between proteins that build the mechanotransducing complex have recently begun to be addressed experimentally.[56,97]

More challenging and most critical, the hypothesis that a degenerin-containing channel is mechanically gated must be addressed. This may be particularly difficult because at present it is not straightforward to record directly from tiny *C. elegans* neurons. Expression of the MEC-4/MEC-10 or (UNC-8/DEL-1) channel in heterologous systems such as *Xenopus* oocytes is complicated by the presence of the many endogenous mechanically gated ion channels,[98–100] and by the likely possibility that not only the multimeric channel, but essential interacting proteins will have to be assembled to gate the channel.[4,53]

A question that remains to be addressed is whether the mammalian counterparts of the *C. elegans* degenerins play specialized roles in mechanical signaling in humans. A significant step toward addressing this question has been accomplished with the demonstration that BNC1 is involved in mechanosensory signaling in the skin as we have described above. Even though, the candidacy of BNC1 for being in the core of a mechanotransducing complex was greatly boosted by these results, a demanding critic would argue that albeit very strong, it still remains just a candidacy. The potential role of BNC1 as part of the core mechanotransducing channel can still only be inferred from these experiments and is not directly proven. It is still possible that BNC1 forms or participates in an auxiliary channel that facilitates the function of the actual mechanotransducing channel. A BNC1 knockout does not completely eliminate the responses to mechanical stimuli.[12] The incomplete nature of the BNC1 deficiency effects indicates that even if BNC1 is indeed part of the core mechanosensory channel, it most likely is not the only critical one. Alternatively, there might be more than one, different mechanotransducing complexes within one neuron, with different properties and composition. The above arguments however, are by no means confined to BNC1. On the same basis, MEC-4/MEC-10 and UNC-8/DEL-1 in *C. elegans* as well as PPK in *Drosophila* might not be parts of the real mechanotransducer but only auxiliary ion channels.

The recent identification of another strong candidate mechanosensory channel, the *Drosophila* NompC, adds to the list of candidate mechanosensitive ion channels.[101] NompC is unrelated in amino acid sequence to DEG/ENaC channels and is required for normal mechanosensitive currents in fly hair bristles.[101] Evidence implicating NompC in mechanotransduction is especially convincing given the supporting electrophysiological analysis that is feasible in this system, and the availability of mutants with altered properties and intermediate effects. Therefore, NompC homologues in other organisms, including humans, emerge as putative mechanosensitive ion channels. Even in this case, however, there are caveats; the absence of NompC does not completely eliminate mechanosensitive currents in *Drosophila* hair bristles. Furthermore, the identities and properties of force-generating tethers of NompC in mechanotransducing complexes will need to be determined. Another issue that needs to be addressed is the potential interplay between DEG/ENaC and NompC channels in mechanosensory cells before a clear understanding of mechanotransduction can be achieved.

Despite the considerable progress that has been achieved on all fronts in recent years toward dissecting the process of sensory mechanotransduction at the molecular level, several thorny questions still beg for answers. What is the gating mechanism of mechanosensitive ion channels? How is tension delivered to the mechanotransducing complex? What additional molecules play a part in the biological response to mechanical stimuli? Are human sensory mechanotransducers similar in composition and function to nematode or *Drosophila* ones? Are DEG/ENaC truly the core mechanosensitive channels or are they merely auxiliary channels/components?

It is important to emphasize that although specialized ion channels most likely comprise the core of every metazoan mechanotransducer, it is the other physically associated proteins that shape its wonderful properties. It is equally important to seek and identify these. Without them, our understanding of mechanical transduction will never be complete even if the identity of the core ion channel is revealed. Let us keep in mind that mechanical sensation at the molecular level in higher organisms is most likely a property of a complex structure involving many components and contacts and not of any single protein.

Several tools could be employed towards this goal, such as yeast two-hybrid screens and biochemical methods of co-purification of channel complexes together with anchoring proteins. The advent of the human genome sequence will provide the full set of testable DEG/ENaC candidates for mechanotransduction in humans. Some of these may be more closely related to nematode proteins specialized for mechanotransduction than currently identified family members, and may be the long-sought human mechanosensors. In addition, fine mutations that do not dramatically incapacitate a candidate channel, might be engineered back into mice to then examine how these correlate to the characteristics of mechanically induced currents.

Characterization of expression patterns of all ASIC and ENaC family members in these animals and genetic knockouts of other candidate mechanotransducer channels will be required to address the question of functional redundancy, work that can be easily pursued in the post-genome era. Obviously such studies should also reveal whether other DEG/ENaC family members are needed for the function of other mechanoreceptors or nociceptors in mouse skin. A question that remains to be resolved is how broadly DEG/ENaC family members will prove to be involved in mechanotransduction. Analyses of the mammalian ENaC channel in lipid bilayers suggests that its gating can be influenced by membrane stretch,[102,103] although interpretation of these studies requires attention to experimental caveats.[104]

A tremendous boost to sensory mechanotransduction studies will be provided when the necessary technology that allows direct recordings from nematode neurons is achieved Although, electrophysiological recording from some *C. elegans* neurons and muscles are possible, the touch receptor neurons and other sensory neurons are beyond the realm of feasibility given the current state of the art.[105,106] The capacity to perform electrophysiological studies on degenerin or other ion channels while they are kept embedded in their natural surroundings is the currently missing tool. Perhaps, the development of new, noninvasive monitoring and measurement technologies will be required in the case of the tiny *C. elegans* neurons.[107,108] Direct, nondestructive recordings from touch receptor neurons coupled with the powerful genetics of *C. elegans* will hopefully allow the complete dissection of a metazoan mechanotransducing complex.

Acknowledgments. We thank our colleagues at IMBB for discussions and comments on the manuscript. We gratefully acknowledge the contributions of numerous investigators that we did not include in this review.

References

1. Koltzenburg, M., Stucky, C.L., and Lewin, G.R., 1997, Receptive properties of mouse sensory neurons innervating hairy skin, *J. Neurophysiol.* **78**: 1841–1850.
2. Blount, P., and Moe, P.C., 1999, Bacterial mechanosensitive channels: integrating physiology, structure and function, *Trends Microbiol* **7**: 420–424.
3. Sukharev, S.I., Blount, P., Martinac, B., and Kung, C., 1997, Mechanosensitive channels of *Escherichia coli*: the MscL gene, protein, and activities, *Annu. Rev. Physiol.* **59**: 633–657.
4. Syntichaki, P., and Tavernarakis, N., 2004, Genetic models of mechanotransduction: The nematode *Caenorhabditis elegans*, *Physiol. Rev.* **84**: 1097–1153.
5. Driscoll, M., and Chalfie, M., 1991, The mec-4 gene is a member of a family of *Caenorhabditis elegans* genes that can mutate to induce neuronal degeneration, *Nature* **349**: 588–593.
6. Huang, M., and Chalfie, M., 1994, Gene interactions affecting mechanosensory transduction in *Caenorhabditis elegans*, *Nature* **367**: 467–470.
7. Liu, J., Schrank, B., and Waterston, R.H., 1996, Interaction between a putative mechanosensory membrane channel and a collagen, *Science* **273**: 361–364.
8. Tavernarakis, N., Shreffler, W., Wang, S., and Driscoll, M., 1997, unc-8, a DEG/ENaC family member, encodes a subunit of a candidate mechanically gated channel that modulates *C. elegans* locomotion, *Neuron* **18**: 107–119.
9. Chalfie, M., and Wolinsky, E., 1990, The identification and suppression of inherited neurodegeneration in *Caenorhabditis elegans*, *Nature* **345**: 410–416.
10. Hummler, E., and Horisberger, J.D., 1999, Genetic disorders of membrane transport. V. The epithelial sodium channel and its implication in human diseases, *Am. J. Physiol.* **276**: G567–571.
11. Kellenberger, S., and Schild, L., 2002, Epithelial sodium channel/degenerin family of ion channels: a variety of functions for a shared structure, *Physiol. Rev.* **82**: 735–767.
12. Price, M.P., Lewin, G.R., McIlwrath, S.L., Cheng, C., Xie, J., Heppenstall, P.A., Stucky, C.L., Mannsfeldt, A.G., Brennan, T.J., Drummond, H.A., et al., 2000, The mammalian sodium channel BNC1 is required for normal touch sensation, *Nature* **407**: 1007–1011.
13. Waldmann, R., and Lazdunski, M., 1998, H(+)-gated cation channels: neuronal acid sensors in the NaC/DEG family of ion channels, *Curr. Opin. Neurobiol.* **8**: 418–424.
14. Lingueglia, E., Champigny, G., Lazdunski, M., and Barbry, P., 1995, Cloning of the amiloride-sensitive FMRFamide peptide-gated sodium channel, *Nature* **378**: 730–733.
15. Adams, C.M., Anderson, M.G., Motto, D.G., Price, M.P., Johnson, W.A., and Welsh, M.J., 1998, Ripped pocket and pickpocket, novel *Drosophila* DEG/ENaC subunits expressed in early development and in mechanosensory neurons, *J. Cell Biol.* **140**: 143–152.

16. Darboux, I., Lingueglia, E., Pauron, D., Barbry, P., and Lazdunski, M., 1998, A new member of the amiloride-sensitive sodium channel family in *Drosophila melanogaster* peripheral nervous system, *Biochem. Biophys. Res. Commun.* **246**: 210–216.
17. Take-Uchi, M., Kawakami, M., Ishihara, T., Amano, T., Kondo, K., and Katsura, I., 1998, An ion channel of the degenerin/epithelial sodium channel superfamily controls the defecation rhythm in *Caenorhabditis elegans*, *Proc. Natl. Acad. Sci. USA* **95**: 11775–11780.
18. Tavernarakis, N., and Driscoll, M., 2001a, Degenerins: At the core of the metazoan mechanotransducer? *Ann. NY Acad. Sci.* **940**: 28–41.
19. Drummond, H.A., Price, M.P., Welsh, M.J., and Abboud, F.M., 1998, A molecular component of the arterial baroreceptor mechanotransducer, *Neuron* **21**: 1435–1441.
20. Price, M.P., McIlwrath, S.L., Xie, J., Cheng, C., Qiao, J., Tarr, D.E., Sluka, K.A., Brennan, T.J., Lewin, G.R., and Welsh, M.J., 2001, The DRASIC cation channel contributes to the detection of cutaneous touch and acid stimuli in mice, *Neuron* **32**: 1071–1083.
21. Sulston, J.E., and Horvitz, H.R., 1977, Post-embriyonic cell lineages of the nematode *Caenorhabditis elegans*, *Dev. Biol.* **56**: 110–156.
22. Sulston, J.E., Schierenberg, E., White, J. G., and Thomson, J.N., 1983, The embryonic cell lineage of the nematode *Caenorhabditis elegans*, *Dev. Biol.* **100**: 64–119.
23. White, J.G., Southgate, E., Thomson, J.N., and Brenner, S., 1976, The structure of the ventral nerve cord of *Caenorhabditis elegans*, *Philos. Trans. R. Soc. Lond. B Biol. Sci.* **275**: 327–348.
24. White, J.G., Southgate, E., Thomson, J.N., and Brenner, S., 1986, The structure of the nervous system of *Caenorhabditis elegans*, *Philos. Trans. R. Soc. Lond. B Biol. Sci.* **314**: 1–340.
25. Francis, R., and Waterston, R.H., 1991, Muscle cell attachment in *Caenorhabditis elegans*, *J. Cell. Biol.* **114**: 465–479.
26. Hresko, M.C., Williams, B.D., and Waterston, R.H., 1994, Assembly of body wall muscle and muscle cell attachment structures in *Caenorhabditis elegans*, *J. Cell Biol.* **124**: 491–506.
27. Walthall, W.W., 1995, Repeating patterns of motoneurons in nematodes: the origin of segmentation? *EXS* **72**: 61–75.
28. Chalfie, M., Sulston, J.E., White, J.G., Southgate, E., Thomson, J.N., and Brenner, S., 1985, The neural circuit for touch sensitivity in *Caenorhabditis elegans*, *J. Neurosci.* **5**: 956–964.
29. Chalfie, M., 1993, Touch receptor development and function in *Caenorhabditis elegans*, *J. Neurobiol.* **24**: 1433–1441.
30. Chalfie, M., 1995, The differentiation and function of the touch receptor neurons of *Caenorhabditis elegans*, *Prog. Brain Res,* **105**: 179–182.
31. Chalfie, M., and Sulston, J., 1981, Developmental genetics of the mechanosensory neurons of *Caenorhabditis elegans*, *Dev. Biol.* **82**: 358–370.
32. Chalfie, M., and Thomson, J.N., 1982, Structural and functional diversity in the neuronal microtubules of *Caenorhabditis elegans*, *J. Cell Biol.* **93**: 15–23.
33. Bargmann, C.I., and Kaplan, J.M., 1998, Signal transduction in the *Caenorhabditis elegans* nervous system, *Annu. Rev. Neurosci.* **21**: 279–308.

34. Colbert, H.A., Smith, T.L., and Bargmann, C.I., 1997, OSM-9, a novel protein with structural similarity to channels, is required for olfaction, mechanosensation, and olfactory adaptation in *Caenorhabditis elegans*, *J. Neurosi.* **17**: 8259–8269.
35. Hart, A.C., Kass, J., Shapiro, J.E., and Kaplan, J.M., 1999, Distinct signaling pathways mediate touch and osmosensory responses in a polymodal sensory neuron, *J. Neurosci.* **19**: 1952–1958.
36. Wicks, S.R., and Rankin, C.H., 1995, Integration of mechanosensory stimuli in *Caenorhabditis elegans*, *J. Neurosci.* **15**: 2434–2444.
37. Mah, K.B., and Rankin, C.H., 1992, An analysis of behavioral plasticity in male *Caenorhabditis elegans*, *Behav. Neural. Biol.* **58**: 211–221.
38. Rankin, C.H., 2002, From gene to identified neuron to behaviour in *Caenorhabditis elegans*, *Nat. Rev. Genet.* **3**: 622–630.
39. Katsura, I., Kondo, K., Amano, T., Ishihara, T., and Kawakami, M., 1994, Isolation, characterization and epistasis of fluoride-resistant mutants of *Caenorhabditis elegans*, *Genetics* **136**: 145–154.
40. The *C. elegans* Sequencing Consortium, 1998, Genome sequence of the nematode *C. elegans*: A platform for investigating biology, *Science* **282**: 2012–2018.
41. Benos, D.J., and Stanton, B.A., 1999, Functional domains within the degenerin/epithelial sodium channel (Deg/ENaC) superfamily of ion channels, *J. Physiol.* **520** Pt 3: 631–644.
42. Tavernarakis, N., and Driscoll, M., 2000, *Caenorhabditis elegans* degenerins and vertebrate ENaC ion channels contain an extracellular domain related to venom neurotoxins, *J. Neurogenet.* **13**: 257–264.
43. Garcia-Anoveros, J., and Corey, D.P., 1997, The molecules of mechanosensation, *Annu. Rev. Neurosci.* **20**: 567–594.
44. Tepass, U., Theres, C., and Knust, E., 1990, crumbs encodes an EGF-like protein expressed on apical membranes of *Drosophila* epithelial cells and required for organization of epithelia, *Cell* **61**: 787–799.
45. Rupp, F., Hoch, W., Campanelli, J.T., Kreiner, T., and Scheller, R.H., 1992, Agrin and the organization of the neuromuscular junction, *Curr. Opin. Neurobiol.* **2**: 88–93.
46. Bevilacqua, M.P., Stengelin, S., Gimbrone, M.A., Jr., and Seed, B., 1989, Endothelial leukocyte adhesion molecule 1: an inducible receptor for neutrophils related to complement regulatory proteins and lectins, *Science* **243**: 1160–1165.
47. Tavernarakis, N., and Driscoll, M., 2001b, Mechanotransduction in *Caenorhabditis elegans*: the role of DEG/ENaC ion channels, *Cell Biochem. Biophys.* **35**: 1–18.
48. Hong, K., and Driscoll, M., 1994, A transmembrane domain of the putative channel subunit MEC-4 influences mechanotransduction and neurodegeneration in *C. elegans*, *Nature* **367**: 470–473.
49. Hong, K., Mano, I., and Driscoll, M., 2000, In vivo structure-function analyses of *Caenorhabditis elegans* MEC-4, a candidate mechanosensory ion channel subunit, *J. Neurosci.* **20**: 2575–2588.
50. Garty, H., and Palmer, L.G., 1997, Epithelial sodium channels: function, structure, and regulation, *Physiol. Rev.* **77**: 359–396.
51. Kellenberger, S., Gautschi, I., and Schild, L., 1999, A single point mutation in the pore region of the epithelial Na+ channel changes ion selectivity by modifying molecular sieving, *Proc. Natl. Acad. Sci. USA* **96**: 4170–4175.

52. Chalfie, M., and Au, M., 1989, Genetic control of differentiation of the *Caenorhabditis elegans* touch receptor neurons, *Science* **243**: 1027–1033.
53. Gu, G., Caldwell, G.A., and Chalfie, M., 1996, Genetic interactions affecting touch sensitivity in *Caenorhabditis elegans*, *Proc. Natl Acad. Sci. USA* **93**: 6577–6582.
54. Tavernarakis, N., and Driscoll, M., 1997, Molecular modeling of mechanotransduction in the nematode *Caenorhabditis elegans*, *Annu. Rev. Physiol.* **59**: 659–689.
55. Ernstrom, G.G., and Chalfie, M., 2002, Genetics of sensory mechanotransduction, *Annu. Rev. Genet.* **36**: 411–453.
56. Goodman, M.B., Ernstrom, G.G., Chelur, D.S., O'Hagan, R., Yao, C.A., and Chalfie, M., 2002, MEC-2 regulates *C. elegans* DEG/ENaC channels needed for mechanosensation, *Nature* **415**: 1039–1042.
57. Syntichaki, P., and Tavernarakis, N., 2003, The biochemistry of neuronal necrosis: rogue biology? *Nat. Rev. Neurosci.* **4**: 672–684.
58. Schild, L., and Kellenberger, S., 2001, Structure function relationships of ENaC and its role in sodium handling, *Adv. Exp. Med. Biol.* **502**: 305–314.
59. Schild, L., Schneeberger, E., Gautschi, I., and Firsov, D., 1997, Identification of amino acid residues in the alpha, beta, and gamma subunits of the epithelial sodium channel (ENaC) involved in amiloride block and ion permeation, *J. Gen. Physiol.* **109**: 15–26.
60. Garcia-Anoveros, J., Garcia, J.A., Liu, J.D., and Corey, D.P., 1998, The nematode degenerin UNC-105 forms ion channels that are activated by degeneration- or hypercontraction-causing mutations, *Neuron* **20**: 1231–1241.
61. Waldmann, R., Champigny, G., and Lazdunski, M., 1995, Functional degenerin-containing chimeras identify residues essential for amiloride-sensitive Na+ channel function, *J. Biol. Chem.* **270**: 11735–11737.
62. Syntichaki, P., and Tavernarakis, N., 2002, Death by necrosis: Uncontrollable catastrophe, or is there order behind the chaos? *EMBO Rep.* **3**: 604–609.
63. Hall, D.H., Gu, G., Garcia-Anoveros, J., Gong, L., Chalfie, M., and Driscoll, M., 1997, Neuropathology of degenerative cell death in *Caenorhabditis elegans*, *J. Neurosci.* **17**: 1033–1045.
64. Park, E.C., and Horvitz, H.R., 1986, Mutations with dominant effects on the behavior and morphology of the nematode *Caenorhabditis elegans*, *Genetics* **113**: 821–852.
65. Shreffler, W., Magardino, T., Shekdar, K., and Wolinsky, E., 1995, The unc-8 and sup-40 genes regulate ion channel function in *Caenorhabditis elegans* motorneurons, *Genetics* **139**: 1261–1272.
66. Shreffler, W., and Wolinsky, E., 1997, Genes controlling ion permeability in both motorneurons and muscle, *Behav. Genet.* **27**: 211–221.
67. Garcia-Anoveros, J., Ma, C., and Chalfie, M., 1995, Regulation of *Caenorhabditis elegans* degenerin proteins by a putative extracellular domain, *Curr. Biol.* **5**: 441–448.
68. Goodman, M.B., and Schwarz, E.M., 2003, Transducing touch in *Caenorhabditis elegans*, *Annu. Rev. Physiol.* **65**: 429–452.
69. Du, H., Gu, G., William, C.M., and Chalfie, M., 1996, Extracellular proteins needed for *C. elegans* mechanosensation, *Neuron* **16**: 183–194.
70. Rajaram, S., Spangler, T.L., Sedensky, M.M., and Morgan, P.G., 1999, A stomatin and a degenerin interact to control anesthetic sensitivity in *Caenorhabditis elegans*, *Genetics* **153**: 1673–1682.

71. Tavernarakis, N., Everett, J.K., Kyrpides, N.C., and Driscoll, M., 2001, Structural and functional features of the intracellular amino terminus of DEG/ENaC ion channels, *Curr. Biol.* **11**: R205–208.
72. Gillespie, P.G., and Walker, R.G., 2001, Molecular basis of mechanosensory transduction, *Nature* **413**: 194–202.
73. Hamill, O.P., and Martinac, B., 2001, Molecular basis of mechanotransduction in living cells, *Physiol. Rev.* **81**: 685–740.
74. Hudspeth, A.J., 1989, How the ear's works work, *Nature* **341**: 397–404.
75. Pickles, J.O., and Corey, D.P., 1992, Mechanoelectrical transduction by hair cells, *Trends Neurosci.* **15**: 254–259.
76. Alvarez de la Rosa, D., Krueger, S.R., Kolar, A., Shao, D., Fitzsimonds, R.M., and Canessa, C.M., 2003, Distribution, subcellular localization, and ontogeny of ASIC1 in the mammalian central nervous system, *J. Physiol.* **546**: 77–87.
77. Wemmie, J.A., Askwith, C.C., Lamani, E., Cassell, M.D., Freeman, J.H., Jr., and Welsh, M.J., 2003, Acid-sensing ion channel 1 is localized in brain regions with high synaptic density and contributes to fear conditioning, *J. Neurosci.* **23**: 5496–5502.
78. Deval, E., Salinas, M., Baron, A., Lingueglia, E., and Lazdunski, M., 2004, ASIC2b-dependent regulation of ASIC3, an essential acid-sensing ion channel subunit in sensory neurons, via the partner protein PICK-1, *J. Biol. Chem.* **279**:19531–19539. Epub 2004 Feb 19.
79. Waldmann, R., Champigny, G., Lingueglia, E., De Weille, J. R., Heurteaux, C., and Lazdunski, M., 1999, H(+)-gated cation channels, *Ann NY Acad. Sci.* **868**: 67–76.
80. Alvarez de la Rosa, D., Zhang, P., Shao, D., White, F., and Canessa, C.M., 2002, Functional implications of the localization and activity of acid-sensitive channels in rat peripheral nervous system, *Proc. Natl. Acad. Sci. USA* **99**: 2326–2331.
81. Reeh, P.W., and Steen, K.H., 1996, Tissue acidosis in nociception and pain, *Prog. Brain Res.* **113**: 143–151.
82. Drummond, H.A., Welsh, M.J., and Abboud, F.M., 2001, ENaC subunits are molecular components of the arterial baroreceptor complex, *Ann NY Acad. Sci.* **940**: 42–47.
83. Drummond, H.A., Abboud, F.M., and Welsh, M.J., 2000, Localization of beta and gamma subunits of ENaC in sensory nerve endings in the rat foot pad, *Brain Res.* **884**: 1–12.
84. Garcia-Anoveros, J., Derfler, B., Neville-Golden, J., Hyman, B.T., and Corey, D.P., 1997, BNaC1 and BNaC2 constitute a new family of human neuronal sodium channels related to degenerins and epithelial sodium channels, *Proc. Natl. Acad. Sci. USA* **94**: 1459–1464.
85. Price, M.P., Snyder, P.M., and Welsh, M.J., 1996, Cloning and expression of a novel human brain Na+ channel, *J. Biol. Chem.* **271**: 7879–7882.
86. Waldmann, R., Champigny, G., Voilley, N., Lauritzen, I., and Lazdunski, M., 1996, The mammalian degenerin MDEG, an amiloride-sensitive cation channel activated by mutations causing neurodegeneration in *Caenorhabditis elegans*, *J. Biol. Chem.* **271**: 10433–10436.
87. Champigny, G., Voilley, N., Waldmann, R., and Lazdunski, M., 1998, Mutations causing neurodegeneration in *Caenorhabditis elegans* drastically alter the pH sensitivity and inactivation of the mammalian H+-gated Na+ channel MDEG1, *J. Biol. Chem.* **273**: 15418–15422.

88. Lingueglia, E., de Weille, J.R., Bassilana, F., Heurteaux, C., Sakai, H., Waldmann, R., and Lazdunski, M., 1997, A modulatory subunit of acid-sensing ion channels in brain and dorsal root ganglion cells, *J. Biol. Chem.* **272**: 29778–29783.
89. Chen, C.C., England, S., Akopian, A.N., and Wood, J.N., 1998, A sensory neuron-specific, proton-gated ion channel, *Proc. Natl. Acad. Sci. USA* **95**: 10240–10245.
90. Garcia-Anoveros, J., Samad, T.A., Zuvela-Jelaska, L., Woolf, C.J., and Corey, D.P., 2001, Transport and localization of the DEG/ENaC ion channel BNaC1alpha to peripheral mechanosensory terminals of dorsal root ganglia neurons, *J. Neurosci.* **21**: 2678–2686.
91. Jarvilehto, T., Hamalainen, H., and Laurinen, P., 1976, Characteristics of single mechanoreceptive fibres innervating hairy skin of the human hand, *Exp. Brain Res.* **25**: 45–61.
92. Jarvilehto, T., Hamalainen, H., and Soininen, K., 1981, Peripheral neural basis of tactile sensations in man: II. Characteristics of human mechanoreceptors in the hairy skin and correlations of their activity with tactile sensations, *Brain Res.* **219**: 13–27.
93. Chalfie, M., 1997, A molecular model for mechanosensation in *Caenorhabditis elegans*, *Biol. Bull.* **192**: 125.
94. Eberl, D.F., Duyk, G.M., and Perrimon, N., 1997, A genetic screen for mutations that disrupt an auditory response in *Drosophila melanogaster*, *Proc. Natl. Acad. Sci. USA* **94**: 14837–14842.
95. Nicolson, T., Rusch, A., Friedrich, R.W., Granato, M., Ruppersberg, J.P., and Nusslein-Volhard, C., 1998, Genetic analysis of vertebrate sensory hair cell mechanosensation: the zebrafish circler mutants, *Neuron* **20**: 271–283.
96. Kernan, M., Cowan, D., and Zuker, C., 1994, Genetic dissection of mechanosensory transduction: mechanoreception-defective mutations of *Drosophila*, *Neuron* **12**: 1195–1206.
97. Chelur, D.S., Ernstrom, G.G., Goodman, M.B., Yao, C.A., Chen, A.F., O'Hagan, R., and Chalfie, M., 2002, The mechanosensory protein MEC-6 is a subunit of the C. elegans touch-cell degenerin channel, *Nature* **in press**.
98. Hamill, O.P., Lane, J.W., and McBride, D.W., Jr., 1992, Amiloride: a molecular probe for mechanosensitive channels, *Trends Pharmacol. Sci.* **13**: 373–376.
99. McBride, D.W., Jr., and Hamill, O.P., 1992, Pressure-clamp: a method for rapid step perturbation of mechanosensitive channels, *Pflügers Arch.* **421**: 606–612.
100. Zhang, Y., Gao, F., Popov, V.L., Wen, J.W., and Hamill, O.P., 2000, Mechanically gated channel activity in cytoskeleton-deficient plasma membrane blebs and vesicles from *Xenopus* oocytes, *J Physiol* **523**(1): 117–130.
101. Walker, R.G., Willingham, A.T., and Zuker, C.S., 2000, A *Drosophila* mechanosensory transduction channel, *Science* **287**: 2229–2234.
102. Awayda, M.S., Ismailov, II, Berdiev, B.K., and Benos, D.J., 1995, A cloned renal epithelial Na+ channel protein displays stretch activation in planar lipid bilayers, *Am. J. Physiol.* **268**: C1450–1459.
103. Ismailov, II, Berdiev, B.K., Shlyonsky, V.G., and Benos, D.J., 1997, Mechanosensitivity of an epithelial Na+ channel in planar lipid bilayers: release from Ca2+ block, *Biophys. J.* **72**: 1182–1192.
104. Rossier, B.C., 1998, Mechanosensitivity of the epithelial sodium channel (ENaC): controversy or pseudocontroversy? *J. Gen. Physiol.* **112**: 95–96.

105. Avery, L., Raizen, D., and Lockery, S., 1995, Electrophysiological methods, *Methods Cell Biol.* **48**: 251–269.
106. Goodman, M.B., Hall, D.H., Avery, L., and Lockery, S.R., 1998, Active currents regulate sensitivity and dynamic range in *C. elegans* neurons, *Neuron* **20**: 763–772.
107. Bouevitch, O., Lewis, A., Pinevsky, I., Wuskell, J.P., and Loew, L.M., 1993, Probing membrane potential with nonlinear optics, *Biophys. J.* **65**: 672–679.
108. Khatchatouriants, A., Lewis, A., Rothman, Z., Loew, L., and Treinin, M., 2000, GFP is a selective nonlinear optical sensor of electrophysiological processes in *Caenorhabditis elegans*, *Biophys. J.* **79**: 2345–2352.

2
ASICs Function as Cardiac Lactic Acid Sensors During Myocardial Ischemia

Christopher J. Benson and Edwin W. McCleskey*

2.1. Introduction

From the point of view of sensation, the heart is a curious organ. Sensory neurons innervate it, but we all hope never to be aware of them. The only conscious sensation they cause is pain and the only trigger for this is ischemia—when the heart gets insufficient oxygen. It is a sensation often felt only in the last minute of life. This raises two fundamental questions: (1) what purpose do these neurons serve besides mediating ischemic pain?; (2) what signal activates them? Lactic acid is an obvious candidate for the signal because it is released by muscle whenever there is insufficient oxygen. However, researchers have argued that lactic acid cannot be a trigger for ischemic pain because metabolic acidosis does not cause chest pain even though it can drop pH to levels equivalent to those that occur during myocardial infarction.

Here, we review data showing that cardiac sensory neurons express extremely high levels of an ion channel that is triggered by acid and is also enhanced by other chemicals that appear during ischemic acidosis, but not during metabolic acidosis. This resurrects consideration of lactic acid as a pain signal and provides a possible molecular target for selectively suppressing the signal.

2.2. Cardiac Afferent Activation During Ischemia

Cardiac sensory (afferent) neurons continuously sense and fire action potentials in response to mechanical and chemical changes within the heart. However, the only time that people seem to be aware of this flow of information is when they feel pain, or *angina pectoris*, an ominous signal to those stricken with heart disease. It occurs during times of myocardial ischemia or infarction, when blood supply to the heart muscle is insufficient to meet its metabolic demands. The usual

*Christopher J. Benson, Department of Internal Medicine, Carver College of Medicine, University of Iowa, Iowa City, Iowa 52242. chris-benson@uiowa.edu; Edwin W. McCleskey, Vollum Institute, Oregon Health and Sciences University, Portland, Oregon 97239. mccleske@ohsu.edu

underlying disease process is the build-up of atherosclerotic plaque within the blood vessels (coronary arteries) feeding the heart, which can cause pain under the following conditions: (1) myocardial *ischemia*, which is usually due to a chronic arterial luminal narrowing, and generally occurs during times of physical exercise or emotional stress when the metabolic requirements of the heart are increased, or (2) myocardial *infarction* (or heart attack), which most often results from the sudden formation of thrombus at the site of a ruptured plaque, leading to complete obstruction of blood flow through the artery and subsequent myocardial cell death. The problem is immense: as life expectancies have prolonged, ischemic heart disease has become the number one cause of death in the world,[1] and approximately 6 million people suffer from chronic angina in the United States alone.[2]

Consequently, most of the research attention on cardiac sensation has focused on the causes of angina. During the first half of the last century, the neuroanatomical pathways of the cardiac sensory system were defined, primarily by efforts to transect the pain pathways in patients debilitated with chronic angina.[3–5] These studies revealed two populations of cardiac sensory neurons: those that follow the sympathetic tracts (*sympathetic afferents*) to their cell bodies located in the upper thoracic dorsal root ganglia (C_8-T_6), and those following the vagal nerve tracts (*vagal afferents*) to their cell bodies located at the base of the skull in the nodose ganglia (Figure 2.1).[5,6] Surgical transection of the sympathetic afferent tracts relieved angina in most patients; however, in a few cases vagal nerve transection was necessary to alleviate pain.[7] Some evidence suggests that vagal afferents preferentially innervate the inferior wall of the heart, whereas the sympathetic afferents are more diffusely distributed,[8] which may, in part, underlie the differential response to various surgical interventions. The sensory nerve terminals within the heart are not as well studied; they are not as dense as those within somatic structures, and are primarily on the epicardial (outside) surface.[9,10]

While the neuroanatomical pathways of cardiac sensation were becoming better understood, researchers began to pursue the underlying mechanisms regarding sensory responses to ischemia. Two questions have been the focus of much work in the field and are yet to be completely answered: (1) what is the stimulus during ischemia that activates cardiac afferents, and (2) what population of cardiac afferents transmits nociceptive information during ischemia? Regarding the second question, it is generally accepted that finely myelinated Aδ- and unmyelinated C-fiber sympathetic cardiac afferents are essential for pain transmission during ischemia. However, some investigators have shown that specific subgroups of sympathetic afferents are particularly sensitive to ischemia,[11,12] suggesting that a select population of nociceptive afferents may be responsible for transmission of pain signals. On the other hand, ischemia increases the intensity of activation of the majority of cardiac sympathetic afferents.[12,13] This raises a classic question in sensory biology: Is pain caused by activation of a particular set of neurons specialized to trigger pain (called nociceptors) or is it caused by intense firing of neurons that also subserve other functions? Nociceptive-specific neurons may be the rule in pain from skin, but muscle pain likely involves neurons whose alternate purpose is to trigger subconscious sensations. The precise population of cardiac afferents that

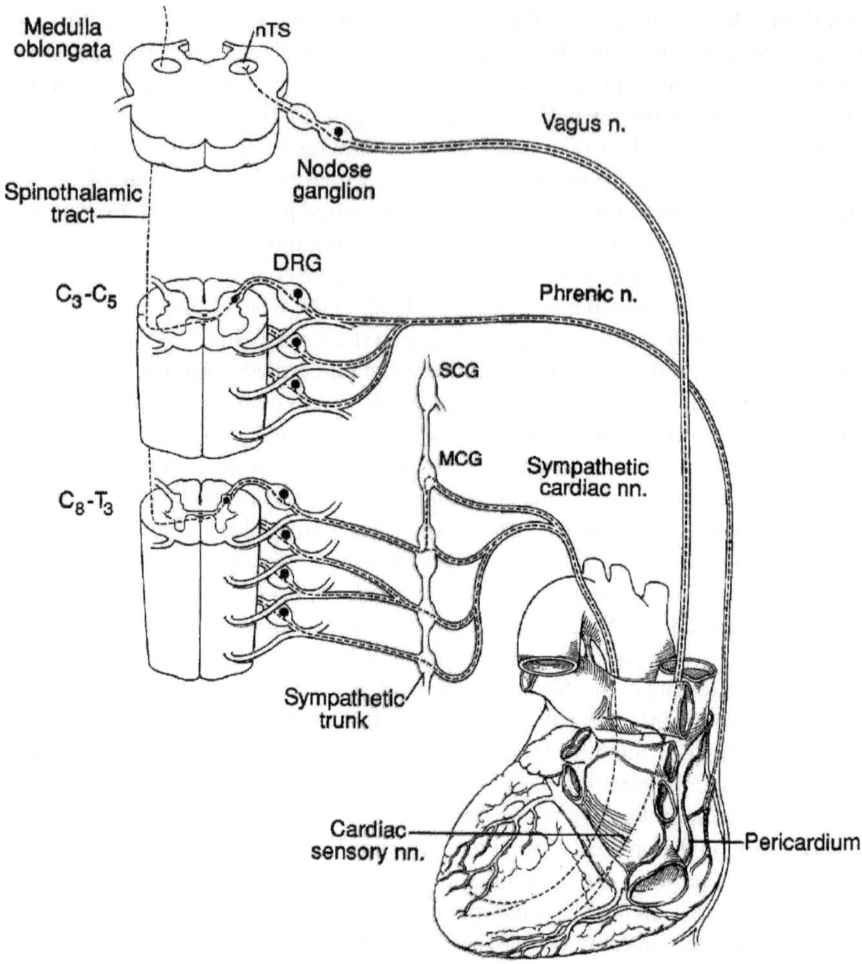

FIGURE 2.1. Sensory innervation of the heart. The myocardium is innervated by sympathetic afferents that follow the sympathetic efferent nerves back to their cell bodies located in the upper thoracic DRG, and cardiac vagal afferents that follow the vagal nerves to the nodose ganglia. Afferent neurons from the pericardium follow the phrenic nerves to their cell bodies in the upper cervical DRG (C_3–C_5). From the respective sensory ganglia, central projections synapse in the spinal cord or brainstem. From Benson et al. (1999).

serve a nociceptive function, and the encoding mechanism for the signal, remain controversial.[14,15]

The first question pertains to the nature of the stimulus that is sensed by cardiac afferents during myocardial ischemia. In the early 1900s a mechanical hypothesis held sway: It was believed that distortion or distention of the ischemic cardiac chambers activated mechanoreceptors on the heart (much like pain generated from other hollow visceral organs)[16]; however, experimental evidence has since proven otherwise. While most cardiac afferents are responsive to mechanical stimuli, this does not seem to correlate with pain responses in animals or humans.[17] Additionally, current clinical practice in cardiology tells us that catheter-based valvular balloon distention, myocardial puncture or biopsy, and radiofrequency-produced burns within the myocardium are all painless procedures in conscious patients.

In the 1930s, Lewis put forth a chemical hypothesis that substances released from ischemic muscle generate pain signals.[18] Since then, it has been generally accepted that sensory activation during myocardial ischemia results from one or more chemical stimuli. Various substances have been implicated, and have been shown to activate cardiac afferent neurons: bradykinin,[19,20] adenosine,[21] serotonin,[22] histamine,[23] ATP,[24] prostaglandins,[25] reactive oxygen species,[11,26] and lactic acid.[27,28] Although incompletely understood, activation of cardiac afferents in the setting of myocardial ischemia probably represents a complex interplay between multiple mediators. Even less is understood regarding the molecular nature of the chemical receptors that transduce these various stimuli into an electrical signal. In this chapter we will focus on our efforts to identify chemical activators of cardiac afferents and the underlying molecular nature of their receptors.

2.3. Acidic Metabolites are Likely Mediators of Sensation During Cardiac Ischemia

The heart is an organ of high metabolic activity and is susceptible to rapid drops in pH during ischemia. Under normal aerobic conditions, the heart readily consumes lactic acid to generate ATP via the respiratory cycle. For example, maximally exercising skeletal muscle generates and releases lactic acid into the circulation. The heart uses this as an energy source: the concentration of lactate within the coronary arteries supplying the heart is generally higher than that in the venous drainage from the heart. However, with insufficient blood supply and oxygen, cardiac myocytes will attempt to maintain contractile function by switching to anaerobic glycolysis. Consequently, lactic acid is generated and accumulates within the cells, which along with the associated drop in pH, inhibits contractile function and contributes to cell death.[29] Myocytes respond by pumping out lactic acid, primarily via a specific lactate transporter, which in turn acidifies the extracellular interstitial spaces within the heart.[30] Additionally, ischemia also contributes to build-up of lactate and other metabolites because low perfusion leads to reduced washout.

What are the concentrations of lactate and H^+ in the heart during ischemia? In isolated ischemic hearts, myocardial intracellular pH drops from about 7.0 to 6.0.[31,32] The extracellular pH, which would be the signal available to trigger sensory neurons, drops within 5 minutes from 7.4 to 7.0. It gets lower only when there is complete loss of blood flow for prolonged times, conditions that cause necrosis.[33,34] Occlusion of coronary blood flow *in vivo* generates a similar drop in pH to the 7.0 range (Figure 2.2A).[28] It is the subtle change—the drop to near

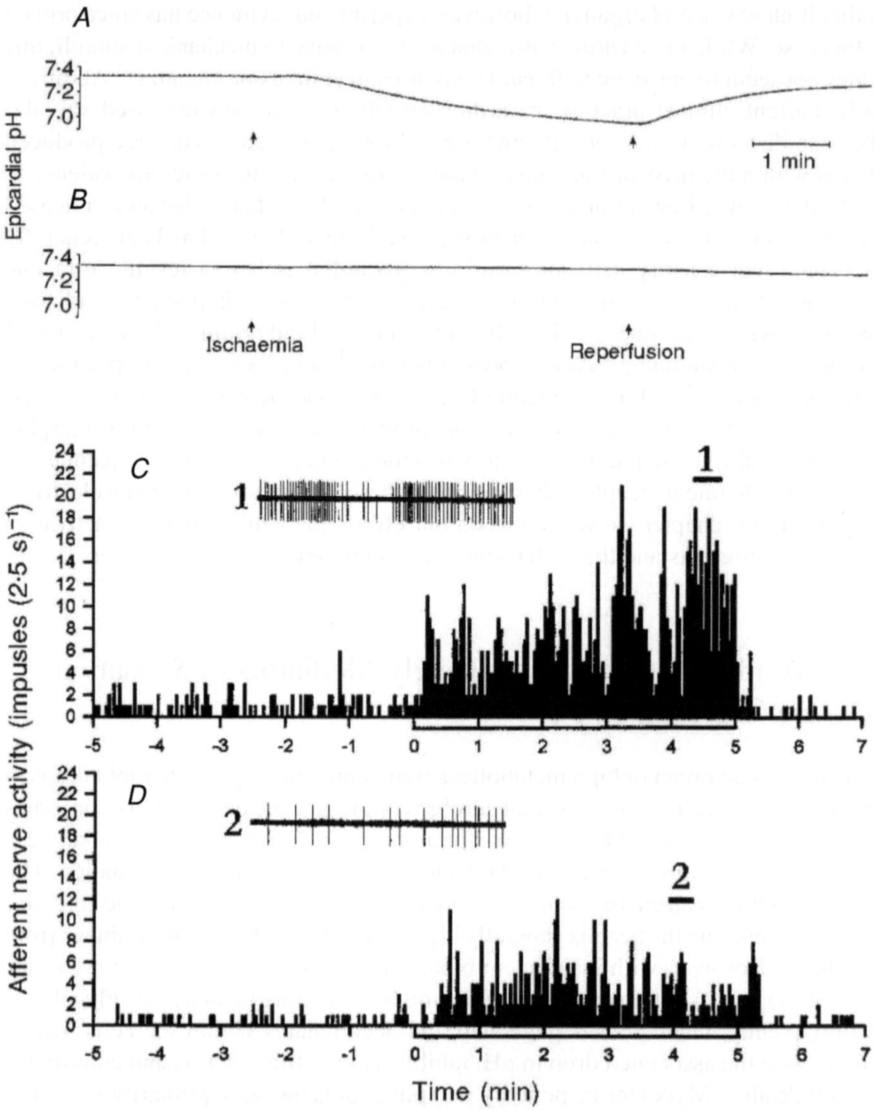

FIGURE 2.2. Myocardial ischemia induces a drop in pH that contributes to cardiac afferent activation. Epicardial pH is lowered during 5 minutes of ischemia (A); this is prevented by infusion of isotonic neutral phosphate buffer into the pericardial sac (B). Frequency histograms of action potentials recorded from a cardiac sympathetic afferent during control, ischemia, and reperfusion before (C) and after (D) pericardial infusion of isotonic neutral phosphate buffer. From Pan et al. (1999).

neutral—that occurs at the time associated with pain.[35] If such a small pH change can be the cause of the pain, there must be a very sensitive detector expressed in cardiac muscle.

Can acidic metabolites associated with ischemia activate cardiac afferents? Uchida and Murao[27] first showed that lactic acid applied to the surface of the heart caused excitation of cardiac sympathetic afferent fibers, although relatively high concentrations were required—correlating with a pH of 4.58. This pH value is below that achieved during myocardial ischemia, and consequently it has been argued that the H^+ concentrations associated with myocardial ischemia are not adequate to activate cardiac afferents and produce pain.[7,36] However, it appears that buffering within interstitial spaces keeps extracellular pH from ever approaching the low value applied to the surface of the tissue. Pan et al.[28] measured the actual pH achieved in the myocardium during acid application by placing a pH-sensitive needle electrode into the myocardium within 1.0–1.5 mm of the surface. They found that a lactic acid concentration of 50 μg/ml (pH 5.42) produced a robust cardiac afferent activation, even though this only produced a drop in measured myocardial pH to 7.0—a pH value readily achieved within minutes of myocardial ischemia.

To evaluate the role of endogenously produced H^+, Uchida and Murao[27] injected sodium bicarbonate to buffer pH and reported a greater than 50% attenuation of cardiac sympathetic afferent activation induced by coronary artery occlusion. Similarly, Pan et al.[28] added a pH buffer into the pericardial sac surrounding the heart to effectively prevent pH changes during ischemia, and they also found afferent activation was inhibited by greater than 50% (Figure 2.2C,D). Thus, the data indicate that acidosis associated with myocardial ischemia is sufficient to excite cardiac afferents. In addition, while several chemicals probably contribute to normal levels of cardiac afferent activation during ischemia, acidic metabolites are a necessary component.

2.4. Isolated Cardiac Afferents Are Activated by Protons

To identify the molecular components that sense myocardial ischemia, we isolated cardiac afferent neurons in culture. The cultivation of sensory neurons has proven to be a useful model to study different sensory modalities; the cell bodies *in vitro* seem to retain the molecular components necessary for sensory transduction at the nerve terminals *in vivo*.[37] To distinguish cardiac from other sensory neurons, we used a fluorescent tracer dye to label cardiac afferents *in vivo* so that they could later be identified in primary dissociated culture (Figure 2.3A,B). Having isolated labeled cardiac afferents, we first applied a variety of chemicals (implicated in cardiac pain) to isolated rat cardiac and non-cardiac (unlabeled) sensory neurons, and measured the resultant ionic currents by whole-cell patch-clamp.[38]

The most important finding of this experiment was that acidic pH evoked large inward currents in almost all cardiac sympathetic afferents (Figure 2.3C-E).

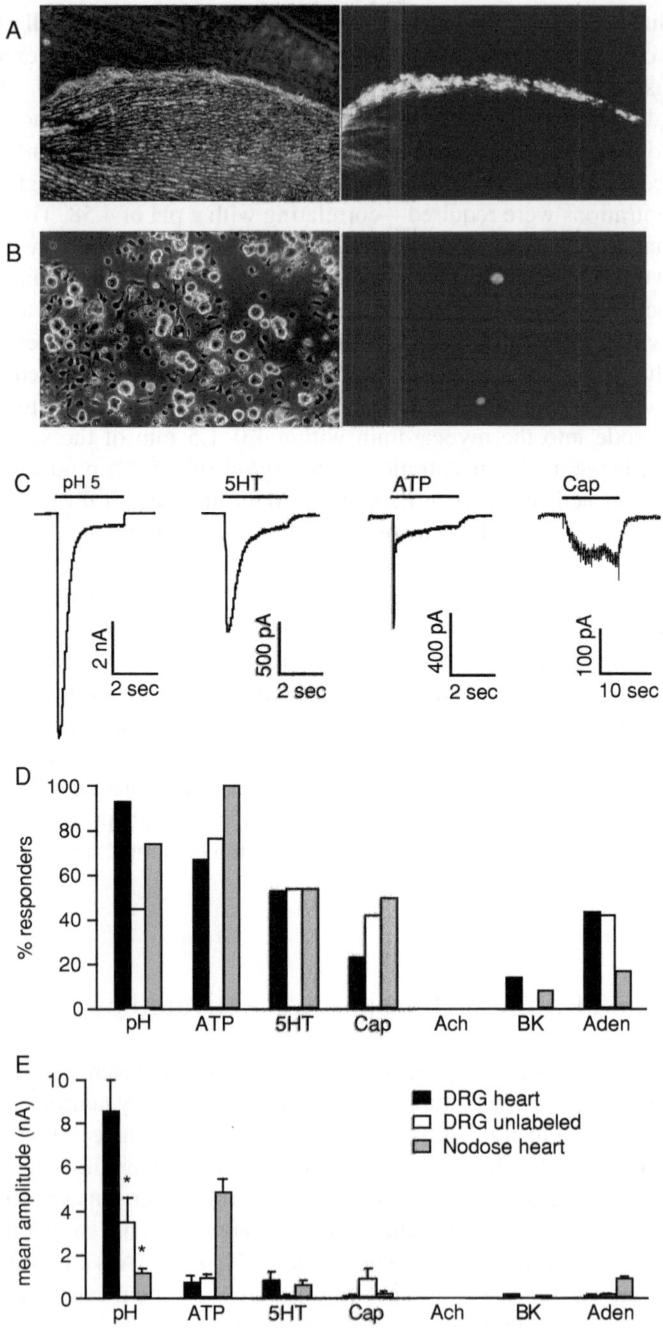
A
B
C
pH 5
5HT
ATP
Cap
2 nA
2 sec
500 pA
2 sec
400 pA
2 sec
100 pA
10 sec
D
% responders
100
80
60
40
20
0
pH
ATP
5HT
Cap
Ach
BK
Aden
E
mean amplitude (nA)
10
8
6
4
2
0
*
*
DRG heart
DRG unlabeled
Nodose heart
pH
ATP
5HT
Cap
Ach
BK
Aden

FIGURE 2.3. Acidic pH activates large currents in isolated cardiac afferents. (A) Corresponding phase (left) and fluorescence (right) micrographs of myocardium 3 weeks after surgical injection of fluorescent tracer dye into the pericardial space. (B) Phase (left) and fluorescence (right) micrographs of two cardiac sympathetic afferents in primary dissociated culture of DRG neurons. (C) Currents evoked by application of indicated agents to cardiac sympathetic afferents. (D) The percentage of cardiac sympathetic (DRG heart), cardiac vagal (nodose heart), and noncardiac (DRG unlabeled) neurons that responded to various agents: [pH, 5.0; ATP, 30 μM; serotonin (5HT), 30 μM; capsaicin (Cap), 1 μM; acetylcholine (ACh), 200 μM; bradykinin (BK), 500 nM; or adenosine (Aden), 200 μM]. (E) Mean amplitudes of the evoked currents of the responding neurons. $^{*}P < .01$ vs. pH-evoked current in DRG heart. From Benson et al. (1999).

Consistent with this, all cardiac sympathetic afferent fibers fire action potentials in response to epicardial application of lactic acid in whole animal models.[12,27] By comparison, a much smaller percentage of noncardiac DRG neurons responded to acid and their currents were significantly smaller. Moreover, the response to other potential chemical mediators generated currents in a lower percentage of cells, and the activated currents were far smaller than those evoked by acid. Thus, while activation of cardiac afferents in the setting of myocardial ischemia most certainly represents a complex interplay between multiple mediators, we have focused on acid and the molecular nature of the pH sensor, as it seems to be expressed at very high levels in cardiac-specific sensory neurons.

2.5. ASICs Are the Proton Sensors in Cardiac Afferents

H^+-gated ion channels were first characterized by Krishtal and co-workers in the early 1980s using electrical recordings of isolated sensory neurons.[39] They describe a channel that opens in response to extracellular acidification, has the unusual characteristic of preferentially passing Na^+ ions through its pore, and is blocked by the diuretic amiloride. Further characterization demonstrated multiple different types of H^+-activated currents, and it became apparent that multiple molecules were involved.[40,41]

In the mid 1990s, two classes of ion channels were cloned that probably account for the bulk of H^+-activated currents described in native neurons. TRPV1 channels are best known for their ability to detect noxious heat and capsaicin, the pungent component of pepper.[42–44] However, they also integrate multiple signals, including voltage, temperature, lipid metabolites, and extracellular acidity.[45–47] At 37°, they are reported to activate at about pH 6.0.[45] While this is much more acidic than that associated with cardiac pain, it is possible that the complex swirl of altered chemistry that accompanies tissue ischemia may increase the acid sensitivity of these molecules.

At the same time, a second class of H^+-gated ion channels was cloned in an effort to identify related members of the DEG/ENaC family of ion channels. This family includes the epithelial Na^+ channel, ENaC, which mediates Na^+ reabsorbtion in the kidneys, lungs, and colon,[48] and the degenerins in *C. elegans*, which participate in mechanosensation.[49] All members in the family are selective for Na^+, and are blocked by amiloride, properties shared by H^+-gated ion channels in sensory neurons. This analogy, along with the fact that several of the newly cloned DEG/ENaC channels were expressed in sensory neurons, led the Lazdunski group to describe the first acid-sensing ion channel (ASIC).[50] We now know three genes within the DEG/ENaC family that encode H^+-gated channels: ASIC1,[50,51] ASIC2,[52] and ASIC3.[53] ASIC1 and ASIC2 both have alternative splice forms involving the amino-termini. Although ASIC4 shows homology, it is not gated by protons.[54,55] We suspect there are no additional ASIC genes; searches of the recently completed mammalian genome sequences have not revealed novel homologous sequences.

Like all DEG/ENaC proteins, ASICs have a large extracellular loop connecting two transmembrane domains, with the amino and carboxyl termini inside the cell. Expression of the ASICs individually in heterologous cells generates transient H^+-gated Na^+ currents (Figure 2.4A). Moreover, when coexpressed in combination, they heteromulterize, producing currents with unique functional properties.[56–58] Expression of the ASICs is restricted to neurons, and mRNA corresponding to each of the subunits is present in sensory neurons.[59–62] Furthermore, ASIC proteins have been detected at nerve terminals,[61–63] where they are poised to transduce sensory stimuli.

With this molecular background in mind, we set out to investigate the identity of the cardiac pH sensor. The biophysical and pharmacological properties of the H^+-evoked currents in cardiac afferents provided the answer. Application of pH less than 7 activated a transient (rapidly activating and desensitizing) current, which was followed by a sustained current only when the pH dropped further, to pH 6 and below (Figure 2.4B). The EC_{50} (pH 6.6) was less acidic than previously reported by other investigators for acid-evoked currents in unselected rat DRG neurons,[64] suggesting that cardiac afferents are particularly sensitive to acidic changes. The transient current was Na^+-selective, and the sustained current was nonselective. Finally, the transient current was inhibited by the amiloride (Figure 2.4C). These properties: the distinct kinetics, exquisite pH sensitivity, Na^+selectivity, and amiloride block, all indicate that H^+-sensing channels in cardiac afferents are ASICs. While our data suggests a minor role of TRPV1 in cardiac sensation (capsaicin generated small amplitude currents in a smaller number of cardiac afferents; Figure 2.3D and E), recent data supports TRPV1 expression in rat cardiac afferents, and a role for TRPV channels in cardiac afferent activation during ischemia.[65,66]

To determine which of the three ASICs contribute to H^+-gated channels in cardiac afferents, we compared the biophysical properties of the native currents to the properties generated by expression of ASIC1 (1a and 1b), ASIC2 (only 2a is expressed in rat sensory neurons), and ASIC3 in heterologous cells.[67] Importantly, the pH sensitivity of ASIC3 most closely matches that of the cardiac afferent channel (Figure 2.4D), and the threshold of activation (pH 7) is well within the

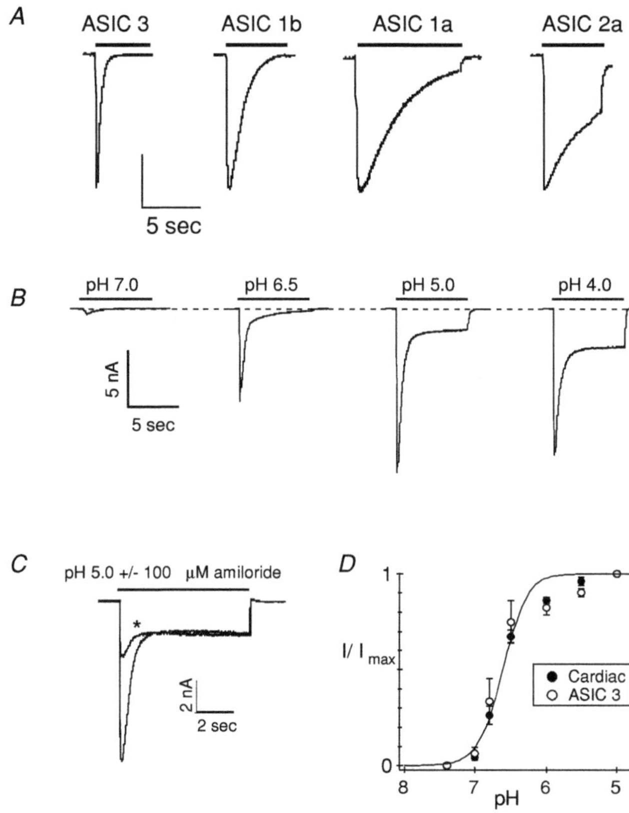

FIGURE 2.4. ASIC3 reproduces the functional properties of the acid-evoked currents in cardiac afferents. (A) Representative acid-evoked currents from COS cells expressing the indicated ASIC subunits. The bars represent a solution change from pH 7.4 to 6, except for ASIC2a, which is evoked by pH 5. (B) Currents evoked by applying various pH solutions to a cardiac sympathetic afferent neuron. (C) Superimposed currents evoked by pH 5.0 and by pH 5.0 plus 100 μM amiloride ([). (D) Average fractional current vs. pH for cardiac afferents (filled circles) and COS-7 cells expressing ASIC3 (open circles). Adapted from Benson et al. (1999), Sutherland et al. (2001), and Benson et al. (2001).

range attained during myocardial ischemia.[28,34] Other properties were also best matched by ASIC3, suggesting it likely is the major constituent of the H^+-gated channel in rat cardiac afferent neurons. However, to match some properties required co-expression of multiple ASIC subunits.[56] For example, we found that co-expression of ASIC3 and ASIC2 reproduced the cation nonselective sustained currents occasionally observed in native neurons. Moreover, the characterization of ASIC channel subunit composition in mice, taking advantage of mice lacking specific ASIC genes, seems to indicate that a majority of ASIC channels in sensory neurons are heteromultimers that consist of ASIC3 in combination with other ASIC subunits.[68]

2.6. ASICs Are Lactate Sensors

It has been observed in whole animal models that lactate is a more potent activator of visceral afferents than H^+ derived from other acid sources.[27] Pan et al.[28] demonstrated that application of lactic acid to the surface of the heart to produce a pH of 7.0 potently activated cardiac afferents. In contrast, application of acidic phosphate buffer or inhalation of CO_2 caused no effect or only slightly increased activity, respectively, despite producing equivalent drops in myocardial pH. Lactic acid is also a more potent stimulator of intestinal and pulmonary afferents.[69,70] This seemingly paradox of lactic acid potency can now be explained by our further understanding of how ASIC channels are activated.

Muscle ischemia causes extracellular lactate to rise to about 15 mM from a resting level below 1 mM.[71,72] Applying 15 mM lactate concentration to isolated cardiac afferents resulted in a ~60% increase in current generated by pH 7 (Figure 2.5). This property was precisely reproduced by applying lactate to heterologously expressed ASIC3. The mechanism involves a shift in the pH sensitivity of the channel, making the channel an even better sensor of the subtle pH changes that occur in the setting of cardiac ischemia. Lactate acts not through a specific binding site, but rather it decreases the concentration of extracellular divalent ions, which are known blockers of ASIC channels.[73] Decreasing extracellular divalent ions can itself open ASIC channels and it potently increases their sensitivity to protons.[74] This unique property of ASICs—to integrate both lactate and H^+—provides a molecular mechanism underlying the observed lactic acid paradox (further supporting a role for ASICs as pH sensors *in vivo*), and makes the channels ideal sensors of the metabolic changes associated with myocardial ischemia.

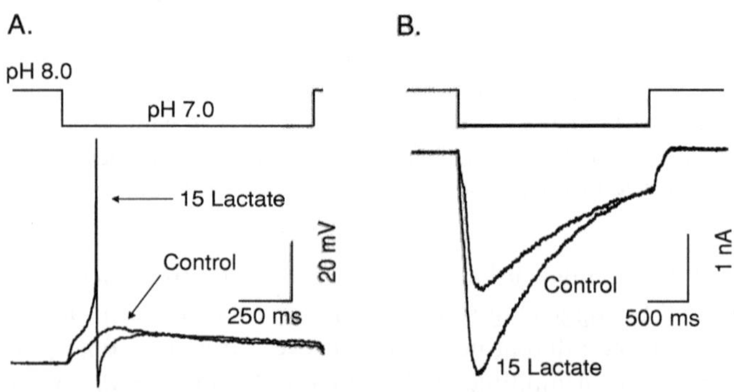

FIGURE 2.5. Lactate potentiates ASICs. Voltage (A) and current (B) recordings from a labeled cardiac sympathetic afferent neuron exposed to pH 7.0 in the presence or absence of 15 mM lactate. The channels are ASICs because the current selectively passed Na^+ and was blocked by 10 μM amiloride (data not shown). Adapted from Immke and McCleskey (2001).

2.7. ASICs May Integrate Multiple Mediators During Ischemia

Multiple chemicals can activate cardiac afferents, and there is some evidence suggesting an additive or synergistic effect. In a rat model of cardiac nociception, Euchner-Wamser et al.[75] found that a mixture of chemical agents led to more avoidance behavior and greater neuronal activation than bradykinin alone. Moreover, in the skin it has been proposed that a combination of chemical mediators produces a more intense sensory activation than any individual mediator alone,[76] and that acid plays a dominant role in this setting.[77]

Recent data suggests that ASICs, in addition to their role as lactate sensors, might integrate multiple chemical signals. Pre-application of a mixture of chemical mediators has been shown to increase H^+-activated ASIC-like currents in sensory neurons.[78] In part, this result is due to transcriptional up-regulation of ASIC expression.[78–80] In addition, some chemicals can increase ASIC current within minutes, suggesting a cellular signaling mechanism.[81] There are a couple of potential signaling mechanisms that might, in part, explain an interaction between ASICs and other agents. First, ASIC2 can be phosphorylated and its function potentiated by protein kinase C (PKC).[82]

Recently, Deval et al.[81] demonstrated that ASIC3 + 2b heteromeric channels (potentially an important ASIC channel in cardiac and other sensory neurons) are positively regulated by a 2-minute pre-application of serotonin or bradykinin via PKC pathway activation. The effect is similar to that produced by lactate: an increase in the pH sensitivity of the channel. Both serotonin and bradykinin can activate PKC via their respective G-protein-coupled receptors, leading to sensitization of sensory neurons and inflammatory hyperalgesia.[83–85] Data suggests ASIC currents are subsequently potentiated by PKC phosphorylation of purported sites on the ASIC2b and –3 subunits.[81] Secondly, ASIC1 and ASIC3 can be phosphorylated by cAMP dependent protein kinase (PKA),[86] although the functional significance is yet unknown. PKA signaling pathways are also important for sensory neuron receptor function.[87,88] Multiple agents that have been implicated in cardiac sensation, including adenosine, serotonin, histamine, and PGE_2, can activate PKA.[89–92] and potentially regulate ASICs.

Evidence suggests multiple chemical mediators may be important to activate cardiac afferents in the setting of ischemia; we hypothesis that lactic acid is a major signal, and ASICs are a major sensor, and that other mediators could, in part, produce effects by modulating ASIC channels.

2.8. Significance

We found that sensory neurons that innervate the heart express high levels of ASIC3 and we showed that it is particularly sensitive to lactic acid at concentrations that occur during cardiac ischemia. If one were to find a magic bullet that selectively blocks ASIC3, what would be its value? And what would be its danger?

Surely, it is dangerous to be unaware of a heart attack, so a complete blocker of cardiac sensation would be problematic. But partial inhibition would provide relief to those who suffer from debilitating angina. Perhaps more importantly, the full physiological and pathophysiological importance of cardiac sensation in the setting of ischemia is missed if the focus is purely on the conscious perception of angina. In fact, the objective measurements of myocardial ischemia often do not correlate well with the presence or severity of angina.[75,93–95] For example, the majority of patients with ischemic heart disease have objectively measurable episodes of ischemia that remain below the threshold of consciousness and are thus, painless—so-called "silent" ischemia.[96] Thus, the conscious perception of chest pain most certainly involves complex central processing and integration at multiple levels, and activation of cardiac sensory neurons is probably necessary but not sufficient to produce pain. Nevertheless, ischemia-induced activation of cardiac afferents, whether painful (conscious) or not, is an important initiator of cardiovascular reflexes in pathologic cardiac conditions.

Cardiac afferents trigger a sympathetic mediated reflex[97,98] that is counterproductive in both acute infarction and in chronic ischemia. The sympatho-excitation contributes to the hemodynamic alterations and frequent arrhythmias encountered in the setting of myocardial ischemia. A detrimental positive feedback loop ensues: Increased sympathetic output leads to increased heart rate, blood pressure, myocardial contractility, and tachyarrhythmias, which in turn further increases metabolic demands on the heart, perpetuating a downward cycle. In fact, most deaths from myocardial infarction occur during the first hour, at a time when this autonomic disturbance is at its most extreme.[99] The remarkable therapeutic benefit provided by beta-receptor blocking agents in the setting of myocardial ischemia and infarction provide further evidence for the pathological importance of this reflex.[100] Understanding the triggers of this reflex (i.e., the molecules that sense the ischemic heart) provides a means to target the sympathetic system at its onset, and to ameliorate its deleterious effects upon cardiac disease.

References

1. *The World Health Report*, 2002. WHO Press.
2. Gibbons, R.J., Abrams, J., Chatterjee, K., Daley, J., Deedwania, P.C., Douglas, J.S., Ferguson, T.B., Jr., Fihn, S.D., Fraker, T.D., Jr., Gardin, J.M., O'Rourke, R.A., Pasternak, R.C., Williams, S.V., Alpert, J.S., Antman, E.M., Hiratzka, L.F., Fuster, V., Faxon, D.P., Gregoratos, G., Jacobs, A.K., and Smith, S.C., Jr., ACC/AHA 2002 guideline update for the management of patients with chronic stable angina—summary article: A report of the American College of Cardiology/American Heart Association Task Force on Practice Guidelines (Committee on the Management of Patients With Chronic Stable Angina), *Circulation*. **107**, 149–58 (2003).
3. Cutler, E.C., Summary of experiences up-to-date in the surgical treatment of angina pectoris., *Am. J. Med. Sci.* **173**, 613–624 (1927).

4. Lingren, I., and Olivecrona, H., Surgical treatment of angina pectoris, *J. Neurosurg.* **4**, 19–39 (1947).
5. White, J.C., Cardiac pain: anatomic pathways and physiologic mechanisms, *Circulation.* **16**, 644–655 (1957).
6. Kuo, D.C., Oravitz, J.J., and DeGroat, W.C., Tracing of afferent and efferent pathways in the left inferior cardiac nerve of the cat using retrograde and transganglionic transport of horseradish peroxidase, *Brain Res.* **321**, 111–118. (1984).
7. Meller, S.T., and Gebhart, G.F., A critical review of the afferent pathways and the potential chemical mediators involved in cardiac pain, *Neuroscience.* **48**, 501–24 (1992).
8. Thames, M.D., Klopfenstein, H.S., Abboud, F.M., Mark, A.L., and Walker, J.L., Preferential distribution of inhibitory cardiac receptors with vagal afferents to the inferoposterior wall of the left ventricle activated during coronary occlusion in the dog, *Circ. Res.* **43**, 512–19 (1978).
9. Miller, M.R., and Kasahara, M., Studies on the nerve endings in the heart, *Am J Anat.* **115**, 217–33 (1964).
10. Barber, M.J., Mueller, T.M., Davies, B.G., and Zipes, D.P., Phenol topically applied to canine left ventricular epicardium interrupts sympathetic but not vagal afferents, *Circ. Res.* **55**, 532–44 (1984).
11. Huang, H.S., Pan, H.L., Stahl, G.L., Longhurst, J.C., Ischemia- and reperfusion-sensitive cardiac sympathetic afferents: influence of H2O2 and hydroxyl radicals, *Am. J. Physiol.* **269**, H888–901 (1995).
12. Pan, H.L., and Chen, S.R., Myocardial ischemia recruits mechanically insensitive cardiac sympathetic afferents in cats, *J. Neurophysiol.* **87**, 660–8 (2002).
13. Huang, M.H., Horackova, M., Negoescu, R.M., Wolf, S., and Armour, J.A., Polysensory response characteristics of dorsal root ganglion neurones that may serve sensory functions during myocardial ischaemia, *Cardiovasc. Res.* **32**, 503–15 (1996).
14. Malliani, A., The elusive link between transient myocardial ischemia and pain, *Circulation.* **73**, 201–4 (1986).
15. Cervero, F., and Janig, W., Visceral nociceptors: a new world order?, *Trends Neurosci.* **15**, 374–8 (1992).
16. Colbeck, E.H., Angina pectoris: a criticism and a hypothesis, *Lancet.* **1**, 793–95 (1903).
17. Davies, G.J., Bencivelli, W., Fragasso, G., Chierchia, S., Crea, F., Crow, J., Crean, P.A., Pratt, T., Morgan, M., and Maseri, A., Sequence and magnitude of ventricular volume changes in painful and painless myocardial ischemia, *Circulation.* **78**, 310–9 (1988).
18. Lewis, T., Pain in muscular ischemia: its relation to anginal pain, *Arch. Int. Med.* **49**, 713–27 (1932).
19. Baker, D.G., Coleridge, H.M., Coleridge, J.C., and Nerdrum, T., Search for a cardiac nociceptor: stimulation by bradykinin of sympathetic afferent nerve endings in the heart of the cat, *J. Physiol. (Lond).* **306**, 519–36 (1980).
20. Lombardi, F., Della Bella, P., Casati, R., and Malliani, A., Effects of intracoronary administration of bradykinin on the impulse activity of afferent sympathetic unmyelinated fibers with left ventricular endings in the cat, *Circ. Res.* **48**, 69–75 (1881).
21. Thames, M.D., Kinugawa, T., and Dibner-Dunlap, M.E., Reflex sympathoexcitation by cardiac sympathetic afferents during myocardial ischemia. Role of adenosine, *Circulation.* **87**, 1698–704 (1993).

22. James, T.N., A cardiogenic hypertensive chemoreflex, *Anesth. Analg.* **69**, 633–46 (1989).
23. Nishi, K., Sakanashi, M., and Takenaka, F., Activation of afferent cardiac sympathetic nerve fibers of the cat by pain-producing substances and by noxious heat, *Pflugers Arch.* **372**, 53–61 (1977).
24. Armour, J.A., Huang, M.H., Pelleg, A., Sylven, C., Responsiveness of in situ canine nodose ganglion afferent neurones to epicardial mechanical or chemical stimuli, *Cardiovasc. Res.* **28**, 1218–25 (1994).
25. Ustinova, E.E., and Schultz, H.D., Activation of cardiac vagal afferents in ischemia and reperfusion: Prostaglandins versus oxygen-derived free radicals, *Circ. Res.* **74**, 904–11 (1994).
26. Ustinova, E.E., and Schultz, H.D., Activation of cardiac vagal afferents by oxygen-derived free radicals in rats, *Circ. Res.* **74**, 895–903 (1994).
27. Uchida, Y., and Murao, S., Acid-induced excitation of afferent cardiac sympathetic nerve fibers, *Am. J. Physiol.* **228**, 27–33 (1975).
28. Pan, H.L., Longhurst, J.C., Eisenach, J.C., and Chen, S.R., Role of protons in activation of cardiac sympathetic C-fibre afferents during ischaemia in cats, *J. Physiol. (Lond).* **518**, 857–66 (1999).
29. Vaughan-Jones, R.D., Eisner, D.A., and Lederer, W.J., Effects of changes of intracellular pH on contraction in sheep cardiac Purkinje fibers, *J. Gen. Physiol.* **89**, 1015–32 (1987).
30. Halestrap, A.P., Wang, X., Poole, R.C., Jackson, V.N., and Price, N.T., Lactate transport in heart in relation to myocardial ischemia, *Am. J. Cardiol.* **80**, 17A–25A (1997).
31. Jacobus, W.E., Taylor, G.J.I., Hollis, D.P., and Nunnally, R.L., Phosphorus nuclear magnetic resonance of perfused working rat hearts, *Nature.* **265**, 756–58 (1977).
32. Yan, G.X., and Kleber, A.G., Changes in extracellular and intracellular pH in ischemic rabbit papillary muscle, *Circ. Res.* **71**, 460–70 (1992).
33. Hirsch, H.J., Franz, C.H.R., Bos, L., Bissig, R., Lang, R., and Schramm, M., Myocardial extracellular K+ and H+ increase and noradrenaline release as possible cause of early arrhythmias following acute coronary artery occlusion in pigs, *J. Mol. Cell Cardiol.* **12**, 579–93 (1980).
34. Cobbe, S.M., and Poole-Wilson, P.A., The time of onset and severity of acidosis in myocardial ischaemia, *J. Mol. Cell Cardiol.* **12**, 745–60 (1980).
35. Remme, W.J., van den Berg, R., Mantel, M., Cox, P.H., van Hoogenhuyze, D.C., Krauss, X.H., Storm, C.J., and Kruyssen, D.A., Temporal relation of changes in regional coronary flow and myocardial lactate and nucleoside metabolism during pacing-induced ischemia, *Am. J. Cardiol.* **58**, 1188–94 (1986).
36. Sylven, C., Angina pectoris: Clinical characteristics, neurophysiological and molecular mechanisms, *Pain.* **36**, 145–67 (1989).
37. Kress, M., and Reeh, P.W., More sensory competence for nociceptive neurons in culture, *Proc. Natl. Acad. Sci. USA.* **93**, 14995–97 (1996).
38. Benson, C.J., Eckert, S.P., and McCleskey, E.W., Acid-evoked currents in cardiac sensory neurons: A possible mediator of myocardial ischemic sensation, *Circ. Res.* **84**, 921–28 (1999).
39. Krishtal, O.A., and Pidoplichko, V.I., Receptor for protons in the membrane of sensory neurons, *Brain Res.* **214**, 150–54 (1981).
40. Krishtal, O.A., and Pidoplichko, V.I., A receptor for protons in the membrane of sensory neurons may participate in nociception, *Neuroscience.* **6**, 2599–601 (1981).

41. Bevan, S., and Yeats, J., Protons activate a cation conductance in a subpopulation of rat dorsal root ganglion neurones, *J. Physiol. (Lond)*. **433**, 145–61 (1991).
42. Caterina, M.J., Schumacher, M.A., Tominaga, M., Rosen, T.A., Levine, J.D., and Julius, D., The capsaicin receptor: a heat-activated ion channel in the pain pathway, *Nature*. **389**, 816–24 (1997).
43. Caterina, M.J., Leffler, A., Malmberg, A.B., Martin, W.J., Trafton, J., Petersen-Zeitz, K.R., Koltzenburg, M., Basbaum, A.I., and Julius, D., Impaired nociception and pain sensation in mice lacking the capsaicin receptor [see comments], *Science*. **288**, 306–13 (2000).
44. Davis, J.B., Gray, J., Gunthorpe, M.J., Hatcher, J.P., Davey, P.T., Overend, P., Harries, M.H., Latcham, J., Clapham, C., Atkinson, K., Hughes, S.A., Rance, K., Grau, E., Harper, A.J., Pugh, P.L., Rogers, D.C., Bingham, S., Randall, A., and Sheardown, S.A., Vanilloid receptor-1 is essential for inflammatory thermal hyperalgesia, *Nature*. **405**, 183–7 (2000).
45. Tominaga, M., Caterina, M.J., Malmberg, A.B., Rosen, T.A., Gilbert, H., Skinner, K., Raumann, B.E., Basbaum, A.I., and Julius, D., The cloned capsaicin receptor integrates multiple pain-producing stimuli [see comments], *Neuron*. **21**, 531–43 (1998).
46. Zygmunt, P.M., Petersson, J., Andersson, D.A., Chuang, H., Sorgard, M., Di Marzo, V., Julius, D., and Hogestatt, E.D., Vanilloid receptors on sensory nerves mediate the vasodilator action of anandamide, *Nature*. **400**, 452–57 (1999).
47. Voets, T., Droogmans, G., Wissenbach, U., Janssens, A., Flockerzi, V., Nilius, B., The principle of temperature-dependent gating in cold- and heat-sensitive TRP channels, *Nature*. **430**, 748–54 (2004).
48. Garty, H., Palmer, L.G., Epithelial sodium channels: function, structure, and regulation, *Physiol. Rev.* **77**, 359–96 (1997).
49. Tavernarakis, N., and Driscoll, M., Molecular modeling of mechanotransduction in the nematode *Caenorhabditis elegans*, *Annu. Rev. Physiol.* **59**, 659–89 (1997).
50. Waldmann, R., Champigny, G., Bassilana, F., Heurteaux, C., and Lazdunski, M., A proton-gated cation channel involved in acid-sensing, *Nature*. **386**, 173–77 (1997).
51. Garcia-Anoveros, J., Derfler, B., Neville-Golden, J., Hyman, B.T., and Corey, D.P., BNaC1 and BNaC2 constitute a new family of human neuronal sodium channels related to degenerins and epithelial sodium channels, *Proc. Natl. Acad. Sci. USA*. **94**, 1459–64 (1997).
52. Price, M.P., Snyder, P.M., and Welsh, M.J., Cloning and expression of a novel human brain Na^+ channel, *J. Biol. Chem.* **271**, 7879–82 (1996).
53. Waldmann, R., Bassilana, F., de Weille, J., Champigny, G., Heurteaux, C., and Lazdunski, M., Molecular cloning of a non-inactivating proton-gated Na^+ channel specific for sensory neurons, *J. Biol. Chem.* **272**, 20975–78 (1997).
54. Akopian, A.N., Chen, C.C., Ding, Y., Cesare, P., and Wood, J.N., A new member of the acid-sensing ion channel family, *Neuroreport*. **11**, 2217–22 (2000).
55. Grunder, S., Geissler, H.S., Bassler, E.L., and Ruppersberg, J.P., A new member of acid-sensing ion channels from pituitary gland, *Neuroreport*. **11**, 1607–11 (2000).
56. Lingueglia, E., de Weille, J.R., Bassilana, F., Heurteaux, C., Sakai, H., Waldmann, R., and Lazdunski, M., A modulatory subunit of acid-sensing ion channels in brain and dorsal root ganglion cells, *J. Biol. Chem.* **272**, 29778–83 (1997).
57. Bassilana, F., Champigny, G., Waldmann, R., de Weille, J.R., Heurteaux, C., and Lazdunski, M., The acid-sensitive ionic channel subunit ASIC and the mammalian

degenerin MDEG form a heteromultimeric H^+-gated Na^+ channel with novel properties, *J. Biol. Chem.* **272**, 28819–22 (1997).

58. Babinski, K., Catarsi, S., Biagini, G., and Seguela, P., Mammalian ASIC2a and ASIC3 subunits co-assemble into heteromeric proton-gated channels sensitive to Gd^{3+}, *J. Biol. Chem.* **275**, 28519–25 (2000).
59. Waldmann, R., and Lazdunski, M., H^+-gated cation channels: neuronal acid sensors in the NaC/DEG family of ion channels, *Curr. Opin. Neurobiol.* **8**, 418–24 (1998).
60. Chen, C.C., England, S., Akopian, A.N., and Wood, J.N., A sensory neuron-specific, proton-gated ion channel, *Proc. Natl. Acad. Sci. USA*. **95**, 10240–45 (1998).
61. Price, M.P., Lewin, G.R., McIlwrath, S.L., Cheng, C., Xie, J., Heppenstall, P.A., Stucky, C.L., Mannsfeldt, A.G., Brennan, T.J., Drummond, H.A., Qiao, J., Benson, C.J., Tarr, D.E., Hrstka, R.F., Yang, B., Williamson, R.A., and Welsh, M.J., The mammalian sodium channel BNC1 is required for normal touch sensation, *Nature*. **407**, 1007–11 (2000).
62. Garcia-Anoveros, J., Samad, T.A., Woolf, C.J., and Corey, D.P., Transport and localization of the DEG/ENaC ion channel BNaC1α to peripheral mechanosensory terminals of dorsal root ganglia neurons, *J. Neurosci.* **21**, 2678–86 (2001).
63. Price, M.P., McIlwrath, S.L., Xie, J., Cheng, C., Qiao, J., Tarr, D.E., Sluka, K.A., Brennan, T.J., Lewin, G.R., and Welsh, M.J., The DRASIC cation channel contributes to the detection of cutaneous touch and acid stimuli in mice, *Neuron*. **32**, 1071–83 (2001).
64. Krishtal, O.A., and Pidoplichko, V.I., A receptor for protons in the nerve cell membrane, *Neuroscience*. **5**, 2325–27 (1980).
65. Zahner, M.R., Li, D.P., Chen, S.R., and Pan, H.L., Cardiac vanilloid receptor 1-expressing afferent nerves and their role in the cardiogenic sympathetic reflex in rats, *J. Physiol.* **551**, 515–23. Epub 2003 Jun 26 (2003).
66. Pan, H.L., and Chen, S.R., Sensing tissue ischemia: another new function for capsaicin receptors?, *Circulation*. **110**, 1826–31. Epub 2004 Sep 13 (2004).
67. Sutherland, S.P., Benson, C.J., Adelman, J.P., and McCleskey, E.W., Acid-sensing ion channel 3 matches the acid-gated current in cardiac ischemia-sensing neurons, *Proc. Natl. Acad. Sci. USA*. **98**, 711–716 (2001).
68. Benson, C.J., Xie, J., Wemmie, J.A., Price, M.P., Henss, J.M., Welsh, M.J., and Snyder, P.M., Heteromultimers of DEG/ENaC subunits form H+-gated channels in mouse sensory neurons, *Proc. Natl. Acad. Sci. USA*. **99**, 2338–43 (2002).
69. Stahl, G.L., and Longhurst, J.C., Ischemically sensitive visceral afferents: importance of H+ derived from lactic acid and hypercapnia, *Am. J. Physiol.* **262**, H748–53 (1992).
70. Hong, J.L., Kwong, K., and Lee, L.Y., Stimulation of pulmonary C fibres by lactic acid in rats: contributions of H+ and lactate ions, *J. Physiol.* **500**, 319–29 (1997).
71. Cohen, R., and Woods, H., Lactic acidosis revisited, *Diabetes*. **32**, 181–91 (1983).
72. Aresta, F., Gerstenblith, G., and Weiss, R.G., Repeated, transient lactate exposure does not "precondition" rat myocardium, *Can. J. Physiol. Pharmacol.* **75**, 1262–66 (1997).
73. Immke, D.C., and McCleskey, E.W., Lactate enhances the acid-sensing Na^+ channel on ischemia-sensing neurons, *Nat. Neurosci.* **4**, 869–70 (2001).
74. Immke, D.C., and McCleskey, E.W., Protons open acid-sensing ion channels by catalyzing relief of Ca2+ blockade, *Neuron*. **37**, 75–84 (2003).
75. Euchner-Wamser, I., Meller, S.T., and Gebhart, G.F., A model of cardiac nociception in chronically instrumented rats: behavioral and electrophysiological effects of pericardial administration of algogenic substances, *Pain*. **58**, 117–28 (1994).

76. Kessler, W., Kirchhoff, C., Reeh, P.W., and Handwerker, H.O., Excitation of cutaneous afferent nerve endings *in vitro* by a combination of inflammatory mediators and conditioning effect of substance P, *Exp. Brain Res.* **91**, 467–76 (1992).
77. Steen, K.H., Steen, A.E., and Reeh, P.W., A dominant role of acid pH in inflammatory excitation and sensitization of nociceptors in rat skin, *in vitro*, *J. Neurosci.* **15**, 3982–89 (1995).
78. Mamet, J., Baron, A., Lazdunski, M., and Voilley, N., Proinflammatory mediators, stimulators of sensory neuron excitability via the expression of acid-sensing ion channels, *J. Neurosci.* **22**, 10662–70 (2002).
79. Voilley, N., de Weille, J., Mamet, J., and Lazdunski, M., Nonsteroid anti-inflammatory drugs inhibit both the activity and the inflammation-induced expression of acid-sensing ion channels in nociceptors, *J. Neurosci.* **21**, 8026–33 (2001).
80. Mamet, J., Lazdunski, M., and Voilley, N., How nerve growth factor drives physiological and inflammatory expressions of acid-sensing ion channel 3 in sensory neurons, *J. Biol. Chem.* **278**, 48907–13. Epub 2003 Sep 30 (2003).
81. Deval, E., Salinas, M., Baron, A., Lingueglia, E., and Lazdunski, M., ASIC2b-dependent regulation of ASIC3, an essential acid-sensing ion channel subunit in sensory neurons via the partner protein PICK-1, *J. Biol. Chem.* **279**, 19531–9. Epub 2004 Feb 19 (2004).
82. Baron, A., Deval, E., Salinas, M., Lingueglia, E., Voilley, N., and Lazdunski, M., Protein kinase C stimulates the acid-sensing ion channel ASIC2a via the PDZ domain-containing protein PICK1, *J. Biol Chem.* **277**, 50463–8. Epub 2002 Oct 23 (2002).
83. Burgess, G.M., Mullaney, I., McNeill, M., Dunn, P.M., and Rang, H.P., Second messengers involved in the mechanism of action of bradykinin in sensory neurons in culture, *J. Neurosci.* **9**, 3314–25 (1989).
84. Barbas, D., DesGroseillers, L., Castellucci, V.F., Carew, T.J., and Marinesco, S., Multiple serotonergic mechanisms contributing to sensitization in aplysia: evidence of diverse serotonin receptor subtypes, *Learn. Mem.* **10**, 373–86 (2003).
85. Nicholson, R., Small, J., Dixon, A.K., Spanswick, D., and Lee, K., Serotonin receptor mRNA expression in rat dorsal root ganglion neurons, *Neurosci. Lett.* **337**, 119–22 (2003).
86. Leonard, A.S., Yermolaieva, O., Hruska-Hageman, A., Askwith, C.C., Price, M.P., Wemmie, J.A., and Welsh, M.J., cAMP-dependent protein kinase phosphorylation of the acid-sensing ion channel-1 regulates its binding to the protein interacting with C-kinase-1, *Proc. Natl. Acad. Sci. USA.* **100**, 2029–34 (2003).
87. Aley, K.O., Levine, J.D., Role of protein kinase A in the maintenance of inflammatory pain, *J. Neurosci.* **19**, 2181–86 (1999).
88. Julius, D., and Basbaum, A.I., Molecular mechanisms of nociception, *Nature.* **413**, 203–10 (2001).
89. Atzori, M., Lau, D., Tansey, E.P., Chow, A., Ozaita, A., Rudy, B., and McBain, C.J., H2 histamine receptor-phosphorylation of Kv3.2 modulates interneuron fast spiking, *Nat. Neurosci.* **3**, 791–98 (2000).
90. England, S., Bevan, S., and Docherty, R.J., PGE2 modulates the tetrodotoxin-resistant sodium current in neonatal rat dorsal root ganglion neurons via the cyclic AMP-protein kinase A cascade, *J. Physiol.* **495**, 429–40 (1996).
91. Taiwo, Y.O., Heller, P.H., and Levine, J.D., Mediation of serotonin hyperalgesia by the cAMP second messenger system, *Neuroscience.* **48**, 479–83 (1992).
92. Taiwo, Y.O., and Levine, J.D., Direct cutaneous hyperalgesia induced by adenosine, *Neuroscience.* **38**, 757–62 (1990).

93. Deanfield, J.E., Shea, M.J., and Selwyn, A.P., Clinical evaluation of transient myocardial ischemia during daily life, *Am. J. Med.* **79**, 18–24 (1985).
94. Malliani, A., Significance of experimental models in assessing the link between myocardial ischemia and pain, *Adv. Cardiol.* **37**, 126–41 (1990).
95. Rosen, S.D., Paulesu, E., Nihoyannopoulos, P., Tousoulis, D., Frackowiak, R.S., Frith, C.D., Jones, T., and Camici, P.G., Silent ischemia as a central problem: regional brain activation compared in silent and painful myocardial ischemia [see comments], *Ann. Intern. Med.* **124**, 939–49 (1996).
96. Schang, S.J., Jr. and Pepine, C.J., Transient asymptomatic S-T segment depression during daily activity, *Am. J. Cardiol.* **39**, 396–402 (1977).
97. Malliani, A., Schwartz, P.J., and Zanchetti, A., A sympathetic reflex elicited by experimental coronary occlusion, *Am. J. Physiol.* **217**, 703–9 (1969).
98. Minisi, A.J., and Thames, M.D., Activation of cardiac sympathetic afferents during coronary occlusion: Evidence for reflex activation of sympathetic nervous system during transmural myocardial ischemia in the dog, *Circulation.* **84**, 357–67 (1991).
99. Webb, S.W., Adgey, A.A., and Pantridge, J.F., Autonomic disturbance at onset of acute myocardial infarction, *Br. Med. J.* **3**, 89–92 (1972).
100. Gottlieb, S.S., McCarter, R.J., and Vogel, R.A., Effect of beta-blockade on mortality among high-risk and low-risk patients after myocardial infarction, *N. Engl. J. Med.* **339**, 489–97 (1998).

3 Molecular Components of Neural Sensory Transduction

DEG/ENaC Proteins in Baro- and Chemoreceptors

François M. Abboud*, Yongjun Lu, and Mark W. Chapleau

Abstract: For several decades, our attempt to characterize afferent signals from cardiovascular or peripheral sensory nerves has focused on the description of action potentials in single fibers classified according to their thickness, their conduction velocity, or the degree of their myelination. We are now at the point where we need to define the molecular components of the ion channels and associated proteins that are responsible for mechano-chemo transduction, nociception, temperature, and touch sensitivity.

In this chapter, we focus on a review of recent studies by several laboratories, including ours, which implicate the DEG/ENaC family of ion channels in mediating the baro- and chemoreceptor responsiveness that is critical to the maintenance of cardiovascular homeostasis.

Members of the family have been identified as sensitive to touch in *C. elegans*, and the evolutionary conservation of homologous members has been traced to *Drosophila* and to mammalians.

Key words: Baroreceptor, DEG/ENaC, ASIC, gene knockout mice, mechanosensitive, proton-sensitive, mechanoelectrical transduction, sympathetic afferent, vagal afferent, nodose ganglion, dorsal root ganglion, non-voltage-gated, amiloride-sensitive.

3.1. Introduction

Sensory nerve endings that are activated by mechanical deformation are present in the carotid sinuses, the aortic arch, and in the heart, lungs, and viscera. They are important regulators of arterial pressure and volume homeostasis.[1–4] The signals originating at nerve terminals travel to the central nervous system (CNS) along the carotid sinus nerve, the aortic depressor nerve, and vagal afferents through petrosal

*The Cardiovascular Research Center and the Departments of Internal Medicine and Molecular Physiology and Biophysics, Carver College of Medicine, University of Iowa, Iowa City, IA 52242, USA; email: francois-abboud@uiowa.edu

and nodose ganglia. Another set of nerve endings in the carotid bodies and aortic bodies are activated by hypoxia, acidosis, and hypercapnia. The signals from these chemoreceptors are also relayed to the CNS along the carotid sinus nerves and vagal afferents. Spinal (sympathetic) afferents sensitive to mechanical stimuli and to inflammatory mediators, cytokines, and protons relay mechano-chemosensations and nociception through dorsal root ganglia. The characterization of the molecular components of ion channels responsible for mechano- and chemotransduction, as well as proton sensitivity, has been the subject of significant recent experimental work. Studies have provided evidence that a candidate family of evolutionary conserved ion channels, the degenerin/epithelial sodium ion channel (DEG/ENaC),[5] is associated with the structure and function of the neuronal sensory mechanotransduction.[6–13] More recently, subunits of these channels have been identified in glomus cells of carotid chemoreceptors.[14]

Members of the DEG/ENaC super family were first discovered in a genetic screen for mechanosensitive genes in nematode *Caenorhabditis elegans*,[13] a primitive worm, which has served as a classical model of molecular genomics. Those channels were also identified in *Drosophila*, in mollusks, and in mammals, including ion channels in epithelial cells (the epithelial sodium channel, ENaC) and proton or acid-sensing ion channels (ASICs). The subunits of ASICs are distributed primarily in brain and peripheral sensory nerves with their soma in the dorsal root ganglia (e.g., peripheral somatic afferents) or in petrosal ganglia (e.g., carotid sinus afferents) and nodose ganglia (e.g., aortic depressor nerves and vagal afferents).

Heteromultimeric constructs from the subunits of ENaC or ASICs, or even from more primitive members, may form the ionic transducers for different types of mechanosensations, nociception, or acid and chemosensitivity. These channels have common functional characteristics that include cation selectivity, voltage independence, and blockade by amiloride or gadolinium.

Based on the work of Guharay and Sachs in the 1980s in skeletal muscle of chick embryos,[15] we identified the mechanosensitive ion channels blocked by gadolinium in the arterial baroreceptors of rabbits[16]and developed an isolated neuronal preparation for mechanosensitive aortic baroreceptor nodose neurons.

We then characterized their electrophysiological properties and, in collaboration with the Welsh laboratory at our institution, began to identify their molecular composition. The work of Chalfie and Driscoll in the *C. elegans*[17] led us to postulate the potential evolutionary conservation of members of the DEG/ENaC family in the mechanosensors of baroreceptors.[7] After a failed attempt at creating a beta ENaC deletion, which proved to be lethal in mice, we resorted to exploration of the functional responses to deletion of ASIC subunits.

The ASIC members of the DEG/ENaC family have been shown to mediate the activation of cardiac spinal (sympathetic) afferents of the DRG neurons during myocardial ischemia. (See Chapter 2 by Benson and McCleskey.) Our recent preliminary work suggests that these proton-sensitive ASIC subunits may be activated in response to acidosis and conceivably might function as chemoreceptive transducers in glomus cells. Although the ASIC designation implies that they may function solely as proton sensors, reports indicate that different heteromultimers of ASIC subunits, as well as ENaC subunits, may contribute also to mechanosensation in various nerve endings. Further, disruption of some of these subunits may result in impairment of a sensory signal or its preservation as a result of compensatory increased expression of remaining subunits.

In the following sections we review our work on the molecular characterization of mechanosensitivity of baroreceptor neurons *in vivo* and *in vitro,* as well as evidence for the role of DEG/ENaC ion channels. We also review different effects of ASIC subunit disruptions on proton and mechanical sensitivities published by various groups of investigators, and discuss their functional implications.

3.2. Baroreceptors and Their Mechanosensitive Ion Channels

Mechanoelectrical transduction of the stretch or deformation of the baroreceptor nerve endings as a result of changes in arterial pressure involves two electrical events (Fig. 3.1A). First, depolarization, often referred to as the receptor potential or generator potential, is proportional to the magnitude of mechanical stimulation. Mechanically gated ion channels that include DEG/ENaC subunits and are permeable to cations (e.g., Na^+ and Ca^{2+}) induce sufficient depolarization to trigger the second electrical event, which is action potential generation.

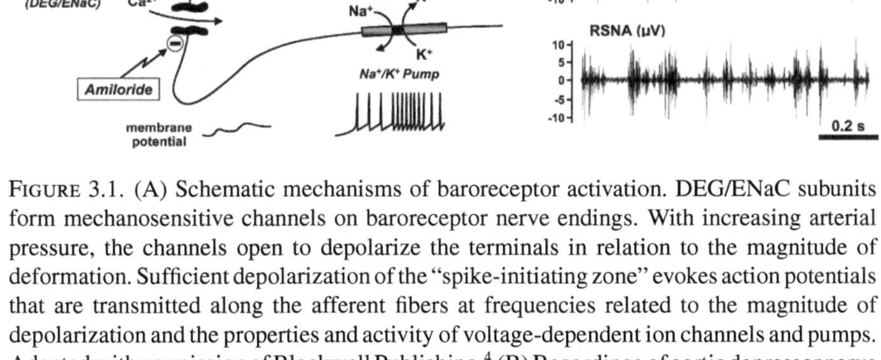

FIGURE 3.1. (A) Schematic mechanisms of baroreceptor activation. DEG/ENaC subunits form mechanosensitive channels on baroreceptor nerve endings. With increasing arterial pressure, the channels open to depolarize the terminals in relation to the magnitude of deformation. Sufficient depolarization of the "spike-initiating zone" evokes action potentials that are transmitted along the afferent fibers at frequencies related to the magnitude of depolarization and the properties and activity of voltage-dependent ion channels and pumps. Adapted with permission of Blackwell Publishing.[4] (B) Recordings of aortic depressor nerve activity (ADNA) and renal sympathetic nerve activity (RSNA) under stimulation of arterial pressure. Bursts of ADNA occur in phase with the arterial pressure pulse, and RSNA exhibits characteristic bursts of activity. Reprinted with permission of the American Physiological Society.[18]

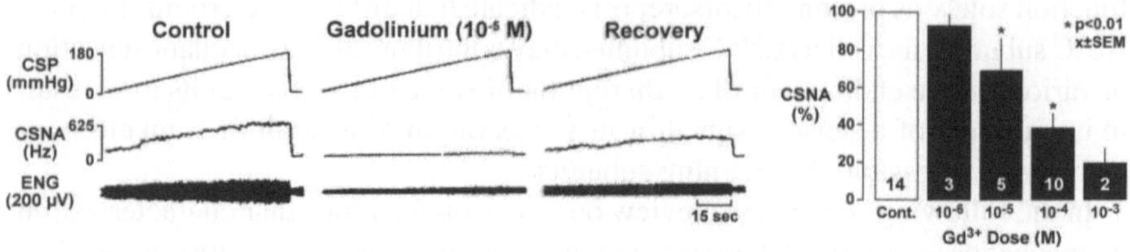

FIGURE 3.2. A ramp increase in carotid sinus pressure (CSP) increased mean carotid sinus nerve activity (CSNA) and the nerve activity seen on the electroneurogram (ENG) in the control condition. Gadolinium (Gd^{3+}) blocked the increase in CSNA during an increasing CSP ramp. After removal of Gd^{3+}, nerve activity was restored. Such nerve activity inhibition is gadolinium dose-dependent (bar graph). Reprinted with permission of the American Society of Clinical Investigation.[16]

These action potentials are initiated by voltage-gated ion channels (Na^+ and K^+) in the "spike-initiating zone" of the nerve terminal and can be modulated by autocrine or paracrine factors that modify the final signal reaching the CNS.

3.2.1. *In Vivo Measurements of Baroreceptor Activity*

In animals, baroreceptor sensitivity can be determined *in vivo* from measurements of changes in aortic depressor nerve activity (ADNA) during changes in arterial pressure and the corresponding reflex changes in, for example, renal sympathetic nerve activity (RSNA) and heart rate.[4] Figure 3.1B shows such an example where ADNA and RSNA recordings from an anesthetized mouse reveal bursts of ADNA activity which are in phase with the arterial pressure pulse, and RSNA exhibits characteristic synchronized bursts of activity.[18] During elevation of arterial pressure there is an increase in ADNA and a reflex suppression of RSNA.

Baroreceptor activity can also be recorded from baroreceptor afferent fibers innervating the isolated carotid sinus. Our first indication that the baroreceptor activity may result from activation of mechanically gated ion channels was our demonstration[16] that gadolinium blocked the increase in carotid sinus nerve activity with the rise in carotid sinus pressure (Fig. 3.2).

The inhibition of nerve activity with gadolinium occurred in a dose-dependent manner and was restored after removal of gadolinium. Because gadolinium had been identified as a relatively selective blocker of mechanosensitive channels, we postulated that the activation of the baroreceptor was dependent on these channels.

3.2.2. *Isolated Baroreceptor Neurons*

The mechanosensitive function of baroreceptors was assessed *in vitro* in isolated neurons in culture by measuring (1) increases in intracellular Ca^{2+} levels using a video microscopic digital imaging of fura-2-AM loaded cells, (2) inward currents in

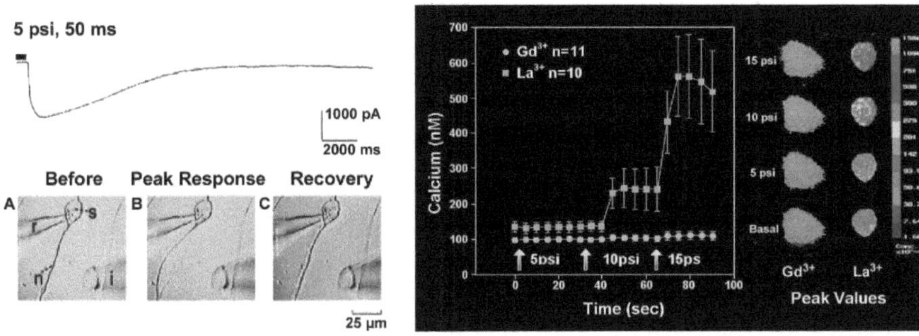

FIGURE 3.3. Effect of mechanical stimulation of neurites on nodose ganglia neurons. Left (top): Typical inward current observed in response to a 5 psi ejection of fluid from a nearby pipette.[19] Left (bottom): photomicrographs taken from a videotape showing the soma of putative baroreceptor neurons (A) before, (B) during, and (C) after the deformation of the neurite. Adapted with permission from Elsevier.[19] Right: Stimulus-dependent increases in intracellular calcium were blocked by gadolinium, but not by lanthanum. Adapted with permission from Lippincott Williams & Wilkins.[20] (See Color Plate 9 in Color Section)

voltage-clamped soma, and (3) resting membrane potential and action potentials in current clamped cells.[7,12,19–23] Mechanical stimulation, for measuring Ca^{2+} signals, was achieved by either probing the neuron with a blunt pipette or by injecting buffer solution from an injection pipette positioned near the soma or the neurite to cause deformation (Fig. 3.3).

There were significant increases in Ca^{2+} levels that were blocked by Gd^{3+} and not by La^{3+}.[20] Both Gd^{3+} and La^{3+} are trivalent lanthenides that block voltage-gated channels, while Gd^{3+} is a specific blocker of mechanosensitive ion channels.[20,21] These results indicate therefore that Ca^{2+} entry was not through voltage-gated Ca^{2+} channels but rather through mechanosensitive ones. There were also significant increases in inward currents during deformation of the neurites (Fig. 3.3).[19]

These currents were resistant to tetrodotoxin, a blocker of the voltage-gated Na^{+} channel, and to tetraethylammonium, 4-aminopyridine, and charybdotoxin, blockers of voltage-gated and Ca^{2+}-activated K^{+} channels.[19,22]

Increases in inward current during the hypoosmotic stretch of the baroreceptor neurons were also blocked by 20 μM Gd^{3+} but not by La^{3+} (20 μM) (Fig. 3.4) or ω-conotoxin (1 μM) (Fig. 3.4B).[23]

3.2.3. *Single Ion Channel*

Figure 3.4C shows the results of cell-attached patch clamp recordings of single-ion channel activity increasing in response to application of negative pressure (−60 mmHg) with little or no channel activity observed when suction was applied to the pipette solution containing 20 μM gadolinium.[22]

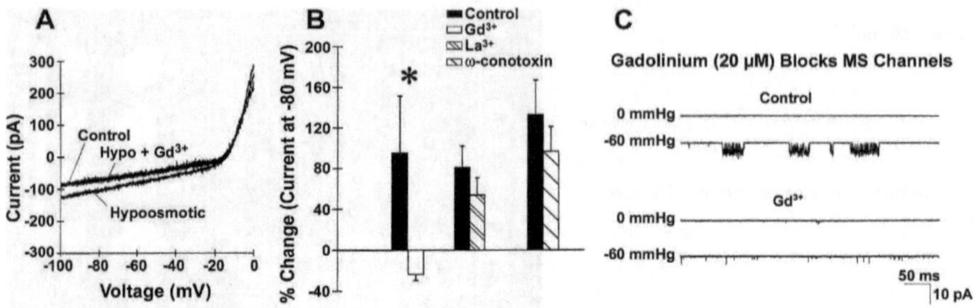

FIGURE 3.4. (A) Current-voltage relation to voltage ramps in a control solution, hypoosmotic solution, and hypoosmotic solution with 20 μM gadolinium from a putative rat aortic baroreceptor neuron; (B) only gadolinium significantly blocked the change in holding current produced by hypoosmotic stretch (*$P < 0.05$). Adapted with permission of Blackwell.[23] (C) Ion channel activity recordings obtained in control condition and with gadolinium in pipette solution under 60 mmHg negative pressure. Adapted with permission of the American Physiological Society.[22]

3.3. Molecular Components of the Mechanoelectrical Transduction Complex

Unlike light or chemically activated receptors or channels, mechanically gated channels are not easily reconstituted, and therefore it has not been possible to identify transduction components by functional screening of mRNA populations.

As mentioned above, DEG/ENaC genes that signalled mechanical responses were identified as a class of ion channels from a genetic screen in *C. elegans*.[9,13,17,24] The cellular phenotype of gain of function mutations resulted in selective degeneration of touch sensory neurons, giving the family the name "degenerin" (DEG). During the identification of the DEG gene family in *C. elegans*, the cDNA encoding the three subunits of the amiloride-sensitive epithelial Na^+ channel (α, β, and γ ENaC) were isolated by expression cloning[25–28] and sequenced. Substantial sequence homology was found between degenerins and ENaC, leading to the identification of additional subfamily members and their characterization by functional expression.[25–31]

3.3.1. Homology Between Members

To date, the DEG/ENaC gene family has seven different branches. The sequences identified that are most relevant to mammalian baroreceptors so far include the subunits encoded by the SCNN1 (sodium channel, non-voltage-gated) genes for the ENaC subunits and the ASIC subfamily encoded by the ACCN (amiloride-sensitive cation channel, neuronal) genes. The DEG/ENaC genes are evolutionarily related with similar gene structures encoding ~30% amino acid sequence homology. The amino acid sequence identity is 15–20% between the different subfamilies, 26–32% within ENaC subfamilies, and 45-60% between the ASIC genes (90% between mouse and human).

The ASIC genes were initially named mammalian degenerins (MDEG) or brain Na^+ channels (BNaC, BNC) due to their expression mainly in the central and peripheral nervous system.[5,32–41] Their activation by protons resulted in their designation as "acid-sensing ion channels" (ASICs).[35,40,41] In a phylogenetic view the DEG and ENaC subfamilies are equally distant from the ASIC subfamily and the name MDEG for ASIC was abandoned.

The ENaC and ASIC genes have been cloned from various species such as human, rat, and mouse, and the related information is available online (http://www.ncbi.nlm.nih.gov). The non-neuronal human α, β, γ, and δENaC genes (SCNN1A, SCNN1B, SCNN1G, and SCNN1D) are located on chromosomes 12, 16, 16, and 1 (human) or 4 (mouse), respectively,[29,30] while the neuronal ASICs (ACCN1, ACCN2, ACCN3, and ACCN4) are on chromosomes 17, 12, 7, and 2, respectively.[32,33,36–43] ENaC subunits show the homology of ~85% between human and rat orthologs, and it is close to 100% for ASIC1, ASIC2, and ASIC4, and ~83% for ASIC3.[5] The highest homology is between specific regions, e.g., the membrane spanning domain and the extracellular cysteine-rich domain.

3.3.2. DEG/ENaC Proteins

DEG/ENaC proteins are relatively small (500–1000 amino acids), and individual subunits share a similar structure,[5,13,44–46] as shown in Figure 3.5, with a large cysteine-rich extracellular loop thought to function as a receptor for extracellular stimuli, two hydrophobic transmembrane domains anchoring the proteins to

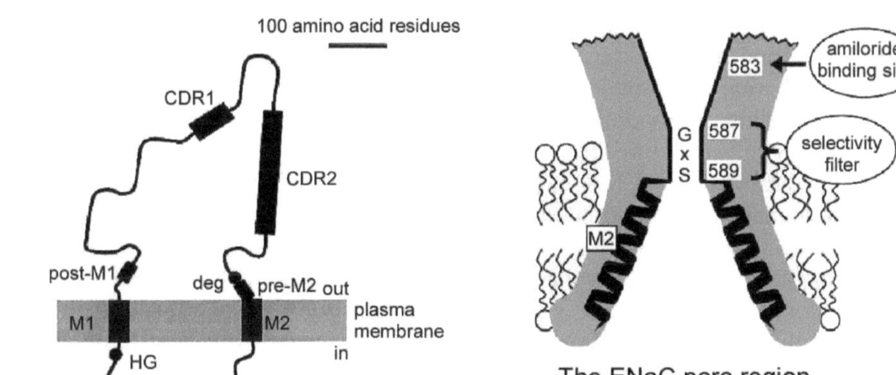

FIGURE 3.5. (A, left) Topological model of an ENaC/DEG subunit. M1, M2, transmembrane segments; HG, His-Gly motif within a N-terminal; post-M1, domain directly downstream of M1; CDR, cysteine-rich domains; pre-M2, hydrophobic domain directly preceding M2; PY, PPPxY domain (5-amino acid sequence with P, Y, and x for proline, tyrosine, and any amino acid, respectively); deg, potential mutation site causing cell degeneration. (B, right) The model of the ENaC pore. G and S denote amino acids glycine and serine, respectively. Adapted with permission of the American Physiological Society.[5]

the membrane, and intracellular amino- and carboxy-termini (N- and C-termini). These channels are highly selective for cations (Na^+ or $Li^+ > K^+ > Ca^{2+}$), and blocked by amiloride. Mutations in the carboxyl region result in channel dysfunction and hypertension in humans.[29,31,47]

These non-voltage-gated ion channels are multimeric, with multiple subunits required to form a functional channel. The number of ENaC subunits is a subject of debate. Some investigators proposed that the ENaC complex is formed by nine subunits, deduced from studies using electrophysiologic, biochemical, and freeze-fracture electron microscopy techniques.[48–50] An alternative model consisting of 4 ENaC subunits (2α, 1β, and 1γ)[44] as well as others[28,34] has been proposed.

ENaC is located in the apical membrane of polarized epithelial cells.[5] The channel is activated by proteolytic cleavage of the extracellular domain of α and γ subunits.[51] This maintains the channel in a constitutively open state that mediates extracellular Na^+ entry into tight epithelia under the driving force of the large electrochemical gradient of Na^+ across the basolateral membrane. Na^+ exit is mediated by Na/K ATPase that maintains a transmembrane Na^+ gradient.[5,31]

3.3.3. Tethering Proteins

A prevailing view describes tethering proteins as obligatory parts of the mechanosensitive DEG/ENaC channel complex with the cytoskeletal and/or extracellular proteins directly tethering and gating sensory mechanotransduction channels. The stress field generated by external loads such as compression and/or tension at the cell surface or induced within the cell is coupled with the extracellular matrix to the receptive endings of mechanoreceptors enabling transmission of force to open the channels.[52,53] The extracellular domains in Figure 3.5 provide such tethering sites for different protein-protein interactions, and even intracellular domain and N- and C-terminals are also involved in such gating processes.[5] Several tethering-like proteins that may play a role in mechanotransduction, include dystrophin, an important link to transmembrane glycoproteins and extracellular collagen,[54] stomatin (MEC-2), a protein linking the channel to cytoskeletal actin,[55,56] c-kinase-1 (PICK-1) binding to ASIC1 protein,[57] and integrin α2β1, a transmembrane protein modulating cutaneous mechanoreceptors.[58]

Other mechanosensitive channels have been identified, e.g., TREK and TRAAK, which are activated by the mechanical stimulation through the bilayer membrane without a necessity to coexpress associated proteins for mechanosensitivity.[59] In other words, depending on the type of ion channels, the tethering protein may or may not be essential for mechanotransduction.

3.4. Neuronal Localization of DEG/ENaC Subunits

The DEG/ENaCs are widely distributed in various tissues including neurons and transporting epithelia and are expressed differentially in a cell specific pattern, indicating a high degree of functional heterogeneity.[5] The subunit stoichiometry may vary in different cell types. In a reconstituted system, the co-expression of

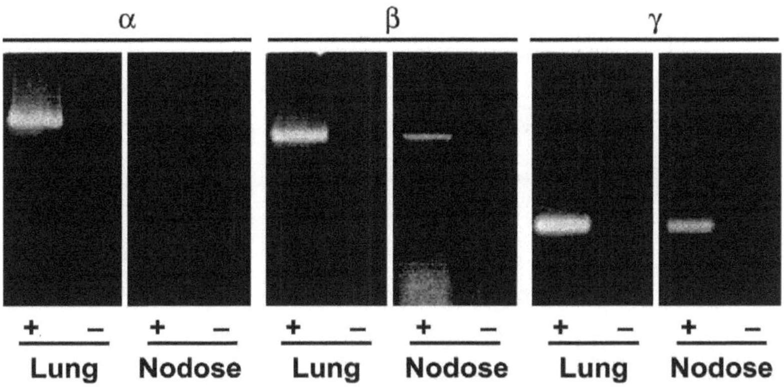

FIGURE 3.6. RT–PCR Analysis of αENaC, βENaC, and γENaC in rat nodose ganglia. Lung, which expresses all three subunits, was used as a control. Reprinted with permission of Elsevier.[7]

αENaC, βENaC, and γENaC subunits leads to maximum channel activity.[34,44] RT-PCR analysis and immunofluorescence were used to determine if mRNA transcripts and proteins for ENaC subunits are expressed in the baroreceptor nerve terminals and aortic nodose neurons of rat. The results show that βENaC and γENaC, but not αENaC, are expressed in nodose neurons (Fig. 3.6).[7,60] The γENaC subunit of the mechanosensitive ion channel is localized at the site of mechanotransduction in baroreceptor afferent nerve terminals (Fig. 3.7).

In addition to ENaC subunits, the mRNA of ASIC subunits has also been localized by RT-PCR both in nodose ganglia and in DRG neurons. Immunofluorescence of ASIC 1, 2, and 3 has been detected in DRG neurons from rat[61] and more recently in nodose neurons and in aortic baroreceptor terminals of the aortic arch of mice.[62] Attempts to quantify the distribution of the various subunits of ASICs have been difficult although work from the Lazdunski laboratory indicates that ASIC3 is overexpressed in the soma of DRG neurons innervating the hind paw of rat as a result of injection of inflammatory mediators in the paw.[63] These studies and other work on acid sensitivity in heterologous systems as well as functional studies in *in vitro* and *in vivo* preparations suggest that ASIC1 and ASIC3 may be more potent as sensors of proton sensitivity than ASIC2 which may predominantly mediate mechanosensitivity.

Recent unpublished work from our laboratory indicates the presence of ASIC3 by RT-PCR and immunofluorescence in glomus cells of rat and is supported by functional studies that demonstrate its role in chemoreceptor activation by acidosis.[14,64]

3.5. ASIC Subunits Act as Acid Sensors and Mechanosensors

Unlike ENaC channels, where all three subunits are necessary for maximal activity,[27] ASICs may form functional channels as heteromultimers when expressed in heterologous cells in combinations.[35,65–67] Because ASIC channels

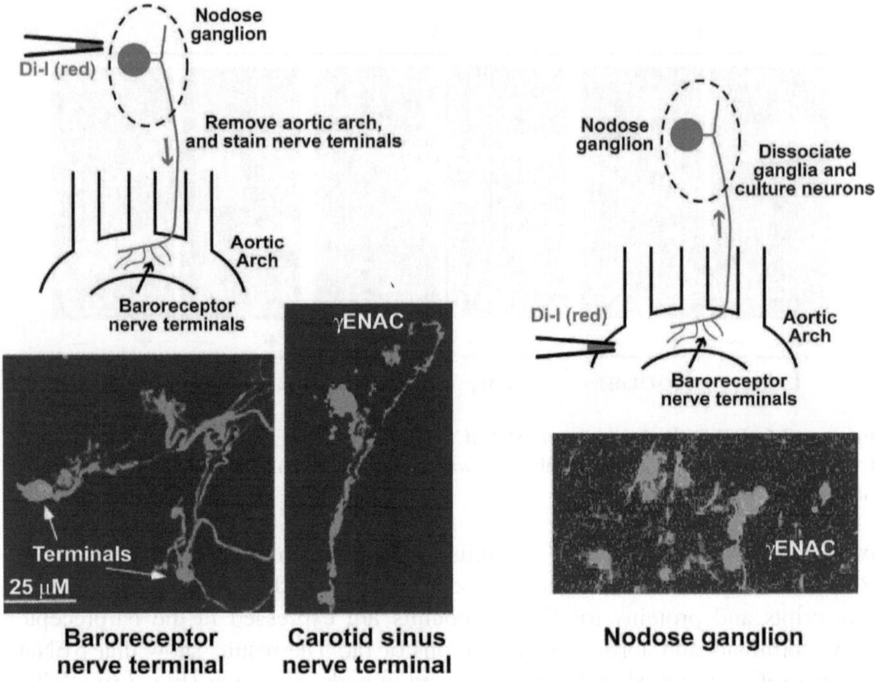

FIGURE 3.7. Immunofluorescence localization of γENaC in aortic baroreceptor nerve terminal, carotid sinus nerve terminal and nodose ganglion neurons of rats using Di-I, a lipophilic tracer dye (1,1′-dioleyl-3,3,3′,3′-tetramethylindocarbocyanine). Reprinted with permission of Elsevier.[7] (See Color Plate 10 in Color Section)

were found predominantly expressed in pain-sensory neurons activated by application of extracellular acidic solution,[5,32] these proton-gated DEG/ENaC subunits had been postulated as pH-sensors in acid-induced nociception that produces pain in humans.

3.5.1. Proton Activation of Sensory Neurons

Protons have been recognized to activate currents in cultured sensory neurons.[6,32,68–70] The currents contain two components: one transient desensitizing and another sustained non-desensitizing.[69–70] The transient currents pass Na^+ preferentially and are inhibited by amiloride, and the sustained component is generally due to opening of nonselective cation channels including vanilloid receptor channels (essential for normal thermal nociception and for thermal hyperalgesia induced by inflammation). The functional composition of the DEG/ENaC channels in medium- to large-sized DRG neurons has been determined. The biophysical properties of transient H^+-gated currents in DRG neurons differ from the properties of the currents in COS-7 cells transfected with a single type of ASIC (ASIC1, 2 or 3), yet co-expression of all three subunits together in COS-7 cells results in currents similar to the endogenous H^+-gated current in DRG neurons.[6]

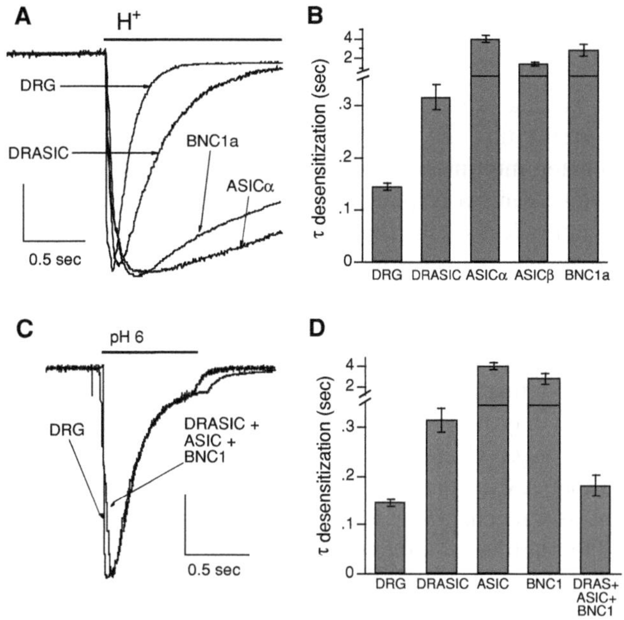

FIGURE 3.8. Desensitization kinetics of H^+-gated currents in DRG neurons are not matched by any one of the DEG/ENaC subunits (A), but are reproduced by co-expression of subunits (C). Mean time constants of desensitization as measured from single exponential fits to the falling phase of the currents for wild-type DRG neurons and COS-7 cells expressing specific subunits are shown in the bar graphs (B and D). Reprinted with permission of the National Academy of Science.[6]

As shown in Figure 3.8, the desensitization kinetics are faster in the DRG current than any of the ASIC channels expressed individually. However, co-expression of the three subunits together produced a channel that desensitized faster than any of the subunits expressed alone, and produced the fast kinetics similar to that of the DRG neurons. The pH sensitivity and recovery from desensitization kinetics were also similar to those of DRG neurons. With the targeted deletion of any one subunit, the two remaining ASIC subunits co-assemble to form a heteromultimeric channel.

ASIC1a (ASIC), ASIC2 (BNC1), and ASIC3 (DRASIC) subunits are also found expressed in large diameter mechanosensory neurons of rat or mouse DRG, i.e., non-nociceptive neurons, generating low-threshold H^+-gated currents in these large cells when studied in culture.[8,11,26,71] ASIC2 has two splice variants, ASIC2a (BNC1a) and ASIC2b (BNC1b); both were detected in wild-type mice but ASIC2b was the predominant form.[8] Paradoxically, protons do not activate the sensory nerve terminals of large-diameter low-threshold mechanoreceptors. Thus, although ASIC channels in non-nociceptive mechanoreceptors can be activated by acidic pH in culture, they may actually function *in vivo* as mechanotransductive channels.

3.5.2. Differences Between Acid Sensitivity of Cardiac DRG and Cardiac Nodose Neurons

Almost all rat cardiac afferents that reside in the DRG possess very large acid-evoked currents that are matched by the properties of ASIC3 alone or ASIC2/ASIC3 heteromultimers.[72,73] In contrast, a smaller percentage of cardiac afferents in nodose ganglia respond to protons, and the resultant currents have much smaller amplitudes (see the chapter by Benson and McCleskey in this book). This selective pattern of activation may explain why the pain associated with myocardial ischemia or infarction (and a significant drop in myocardial pH) is felt to be mediated more by the spinal afferents of the DRG neurons rather than the vagal afferents of the nodose ganglia.[74] Because the sensitivity of DRG is significantly different from that of nodose neurons, it is postulated that DRG connected to spinal afferents may be primarily chemosensitive, whereas the nodose neurons connected to vagal afferents may be predominantly mechanosensitive. We have proposed that ASIC2 plays a more predominant functional role in cardiac nodose neurons as a mechanosensor of volume and pressure changes within the heart mediated through vagal afferents. Cardiac vagal afferents may, however, be chemically activated in the setting of heart failure,[75] and our work on isolated nodose neurons indicates that a subset of mechanosensitive neurons may be also chemosensitive, i.e., bimodal.[76] This chemical activation may compensate for the decreased mechanosensitivity of the cardiac afferents in heart failure and sustain a beneficial reflex sympathoinhibition.[75,76]

3.5.3. Proton Activation of Glomus Cells

The recognized proton sensitivity of ASIC 1 and ASIC3 in particular led us to evaluate the effect of pH on ion channel activation in the carotid body of rat. In current clamp and voltage clamp preparations of glomus cell type 1 cells known to mediate chemoreceptor activity, we found that fast depolarizations that coincide with rapid transient Na^+ current were often observed with gradual step reductions in pH between 7.4 and 6.0.[64] These were blocked by amiloride and not by tetrodotoxin.[64] Along with the demonstration of ASICs by RT-PCR and immunofluorescence in rat carotid body,[14] the results, though preliminary, support a role for ASICs in chemoreceptor activation.

3.6. Functional Responses Confirming the Role of ENaC and ASIC Subunits in Mechano- and Chemosensitivity

Both pharmacologic as well as gene disruption studies have supported the role of ENaC and ASIC subunits in sensory signaling.

3.6.1. Amiloride Sensitivity

A characteristic property of DEG/ENaC-ASIC channels—in addition to the facts that they have large extracellular domains, are non-voltage-gated and non-ligand-gated, cationic, mechanosensitive and proton-sensitive, and with predominant epithelial and neuronal distributions—is their high sensitivity to inhibition by amiloride.

3.6.1.1. Isolated Baroreceptor Neurons

The increases in intracellular Ca^{2+} concentration with mechanical stimulation were selectively blocked by Gd^{3+} and not by La^{3+}, which suggested that calcium entry is through mechanosensitive rather than through voltage-gated channels. Here we show also that the Ca^{2+} transient in mechanically stimulated neurons is also blocked by 100 nM amiloride (Fig. 3.9A), supporting the conclusion that the ENaC subunits that were identified in the baroreceptor nerve terminals are components of the mechanosensitive cationic non-voltage-gated channels.

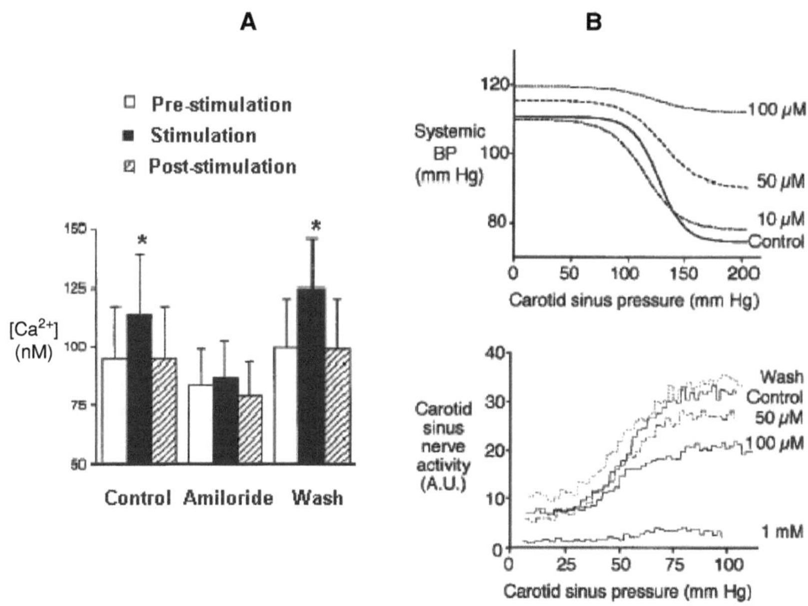

FIGURE 3.9. Effects of amiloride/benzamil on baroreceptors. (A) Bar graph: Mechanically activated Ca^{2+} transients in cultured baroreceptor neurons with puffing buffer solution are blocked by amiloride. (B) Falls in systemic blood pressure (upper graph) and increases in carotid sinus nerve activity (lower graph) in response to increases in carotid sinus pressure are inhibited by increases in concentration of the amiloride analogue benzamil (10 μM to 1 mM). Reprinted with permission of Elsevier.[7]

3.6.1.2. Carotid Sinus Nerve Activity

Further confirmation of the role of DEG/ENaC subunits in transducing baroreceptor activity was obtained in the isolated carotid sinus preparation in rabbits (Fig. 3.9B) where increases in carotid sinus nerve activity with increases in carotid sinus distending pressure and the corresponding reflex decrease in systemic blood pressure were also blocked by benzamil, the amiloride analogue.[7,60]

3.6.2. DEG/ENaC Gene Disruption

3.6.2.1. ENaC Null Mice

Targeted disruption of DEG/ENaC subunits has been attempted. Unfortunately, αENaC$^{-/-}$ mice died from pulmonary edema shortly after birth[77] and βENaC$^{-/-}$ and γENaC$^{-/-}$ mice survived only a few days and died with a significant hyponatremia and hyperkalemia.[78,79] The ENaC is considered to be a multimeric complex composed of α, β, and γ subunits, with all three subunits required for normal channel function.

Rescue of the expression of βENaC in the kidney was attempted using a human βENaC cDNA with the aquaporin-2 promoter of 14 kb engineered to express a kidney-specific gene in transgenic mice. Two out of five transgenic lines had successfully expressed the human βENaC transgene in the kidney but not in neuronal tissues. Further effort is necessary to complete the rescue strategy. Heterozygote βENaC$^{+/-}$ mice are available, but preliminary studies do not reveal any impairment of baroreflexes.

3.6.2.2. ASIC Null Mice

ASIC knockout mice have normal growth, size, temperature, fertility, and life span, but appear to be less active than wild-type. Phenotypes of mice deficient in ASIC1, 2, and 3 have been studied.

Quantitative methods using immunocytochemistry, *in situ* hybridization detection, and RT-PCR analysis have been established and carried out on nodose ganglia, DRG, carotid bodies and nerve terminals[8,11,14,62] to confirm the subunit disruptions.

3.6.2.3. Mechanosensation of Aortic Baroreceptor Neurons in ASIC2 (BNC1) Deficient Mice

The anesthetized mice were given infusions of sodium nitroprusside and phenylephrine intravenously while recording the aortic depressor nerve activity (ADNA) as well as arterial pressure. In preliminary results comparing the wild-type to ASIC2$^{-/-}$ mice, the latter had a more pronounced baroreceptor adaptation with a decline on baroreceptor activity to 50% of the activity level seen in the wild-type.[80] Although this was a significant impairment, the baroreceptor responsiveness to the initial rapid rise in pressure was not suppressed in the null mouse.

More recent findings in telemetered awake ASIC2$^{-/-}$ mice revealed evidence of a significantly impaired baroreflex and mild hypertension.[81]

3.6.2.4. Baroreflex and Chemoreflex Activation During Carotid Artery Occlusion in Mice with ASIC Disruptions

In very recent functional studies, the pressor response to bilateral carotid occlusion (BCO) in mice was used to characterize arterial baro- and chemoreceptor reflexes in mice. After section of the aortic depressor nerves and bilateral vagotomy, BCO in anesthetized mice triggered a significant reproducible hypertensive response that was abrogated by carotid sinus denervation. The reflex was partially reduced by 100% oxygen ventilation, an effect attributed to the elimination of a chemoreceptor hypoxia-driven component of the reflex.[82] This provided an approach for quantitative analysis of baro- and chemoreflex components in ASIC1$^{-/-}$, ASIC2$^{-/-}$, ASIC3$^{-/-}$, and ASIC123$^{-/-}$ mice.[83]

The preliminary results suggest that: (1) ASIC2 provides a major contribution to baroreceptor mechanoelectrical transduction. (2) ASIC1 and ASIC3 contribute to chemoreceptor reflex activation during BCO. (3) Disruption of the ASIC2 subunit may result in a reciprocally augmented expression of the chemosensory ASIC1,3 subunits.

3.6.2.5. Cutaneous Mechanosensation in ASIC2$^{-/-}$ and ASIC3$^{-/-}$ Mice

The DRG neurons of the ASIC2 null mouse have a significant impairment of the low-threshold rapidly adapting mechanoreceptors (RAM) in response to skin displacement (5–20 μm).[8] They also showed some decrease in sensitivity of the slow adapting mechanoreceptors (SAM) compared to the wild-type.[8] However, other slowly conducting D hair afferents and myelinated and nonmyelinated nociceptors were not altered. The green immunostaining for ASIC2 was seen in the lanceolate neural network innervating the hair follicle of ASIC2$^{+/+}$ mice, but not in ASIC2$^{-/-}$ mice.[8,10]

In contrast, the ASIC3 null mouse had an increase in activity of RAM and a decrease in the response of SAM.[11] Other studies have failed to demonstrate a role of ASIC2 or ASIC3 in mechanosensation of DRG neurons.[84]

The results highlight the importance of further examining the role of ASICs in mechanotransduction and support the notion of a heteromultimeric arrangement with some redundancy of effects of the various subunits or interactions that could be either occlusive or facilitatory.

3.7. Studies Negating the Role of ASICs in Mechanosensation

It is interesting to notice that whether the neuronal cells show an alteration of mechanosensitivity after ASICs knockout depends on the target neurons studied.

A change in mechanosensitivity of cutaneous nerves was detected in ASIC2 and ASIC3 null mutants compared to the wild-type.[8,11] However, no change in mechanosensitivity was detected in primary cultured neurons from the ASIC2 knockout compared to wild-type mice where the culture time used in the experiments was unusually long (7-20 days).[85] It is unclear whether the negative results might have been influenced by the prolonged *in vitro* culture time in a medium enriched with nerve growth factors and fetal bovine serum. There is a possibility that other DEG/ENaC channels (not necessarily the ASIC family but ENaC as well) might be expressed in the cultured neurons, restoring the responses that were suppressed as a result of the deleted gene. This view is supported by the observation of Hildebrand et al. who found that the ASIC3 null mouse appeared to develop hearing loss early in life and recover normal hearing at 2 months of age,[86] indicating that the absence of ASIC3 channels in knockout mice might be compensated for by the expression of other mechanosensory genes over time.

The results obtained in the measurements of hearing threshold and noise sensitivity of wild-type and ASIC2 null mice also implied that ASIC2 was not involved in the mechanotransduction of the mammalian cochlea though it might contribute to suprathreshold functions of the cochlea.[87] The negative studies suggest that ASIC2 may not be the only mechanosensory ion channel important for touch mechanosensation, hearing, and visceral mechanoception. Furthermore, different sensory endings may utilize different members of the same or different families with variable degrees of mechano- and chemosensitivity. For instance, cardiac DRG neurons may be more acid sensitive, expressing ASIC3 predominantly, whereas cardiac vagal nodose neurons may be more mechanosensitive. The role of ASIC2 in mechanosensation may be site specific. It is intriguing to speculate that there may be a certain degree of plasticity in the various ASIC subunits that allows restitution of a phenotype when one subunit is deleted.

3.8. Modulation of the Mechanosensitive Complex by Chemical Factors or Hormones

Several paracrine and autocrine factors as well as circulating hormones modulate baroreceptor nerve activity.[4] Besides the indirect vascular effects that could alter the strain on the nerve terminal, neuronal mechanisms can contribute to the modulation. These include the mechanically gated DEG/ENaC depolarization, the voltage-gated spike initiating ion channels, and the tethering proteins that transfer the strain to the mechanosensitive channel.

3.8.1. *Effects of Prostacyclin and Phalloidin*

We observed in the isolated carotid sinus that prostacyclin enhances baroreceptor activity, and we tested whether the excitatory influence was caused by an enhanced mechanosensitive depolarization or increased excitability and action potential generation. The results revealed the latter response without a significant increase in the depolarizing potential with prostacyclin.[76,88] The increased excitability was mediated through inhibition of $IKCa^{2+}$.[89,90] Similarly, the suppressed baroreceptor activity with nitric oxide and reactive oxygen appears to result from nitrosylation of voltage-gated Na^+ channels and methionine oxidation of K^+ channels, respectively, without a change in mechanically-gated depolarization.[4,91–93]

In contrast, phalloidin, which stiffens actin, blocks the mechanosensitive depolarizing potential without altering the action potential response to current injections.[19,90]

These findings reinforce the separate yet complimentary roles of the DEG/ENaC depolarizing channels and the action potential generating voltage-gated channels. Furthermore, the dependence of the mechanosensitive channels on the tethering proteins in cytoskeleton (i.e., actin) and possibly extracellular matrix to transmit the force necessary for opening of the channel is important.

3.9. Perspectives and Conclusions

The simple assumption that the mechanosensory molecule is itself a channel is accepted in the field but has not been definitively proven.[5] Recent results show that ENaC and ASIC subunits of DEG/ENaC are involved in the native baroreceptor complex. Proof for the specific DEG/ENaC subunits that are components of the native baroreceptor complex, and whether they function as homomultimers or heteromultimers, is incomplete as yet. Many basic issues still remain unsolved. Is this putative flexible channel protein affected directly by blood pressure or through an intermediate state to open the closed channel? What is the pressure profile in the membrane under the tension and deformation and how does it interact with charged and polar residues of the DEG/ENaC transmembrane domain? How does the channel protein interact with the membrane lipid?

Recently, cardio-vagal efferent activity has become a major target for diagnosis and intervention in heart disease. Activation of cardiac spinal (sympathetic) afferents of DRG neurons increases sympathetic outflow, whereas the activation of vagal afferents of nodose ganglia causes sympatho-inhibition. The majority of cardiac DRG neurons that receive input from spinal (sympathetic) terminals are very sensitive to low pH, and the nodose neurons that receive cardiac vagal afferents reveal little proton sensitivity and are mechanosensitive. The expression of acid-sensitive DEG/ENaC channels may tend to increase with inflammation. It is challenging to tease out the pathophysiological contributions of these candidate sensory molecules to the functional consequences of excessive activation of sympathetic efferents in heart failure.

Further, defining the differences in the molecular composition of the DEG/ENaC channels between nodose neurons and DRG neurons is necessary to account for the difference in physiological responses. Animal disease models of myocardial ischemia and heart failure have been established for further study,[94] and gene transfer with dominant negative genes specifically and locally to the aortic arch or the carotid sinus provides a useful approach for the further study of the molecular identification of the baroreceptor *in-vivo*.[4] Viral vectors have been successfully used for gene delivery to peripheral targets and brain.[95–97] A particular challenge has been the difficulty in heterologous expression of the mechanosensitive complex from various subunits as has been done with proton sensitive ASIC subunits. The use of antisense oligonucleotides or siRNA to silence gene expression,[98–99] or the use of negative protein mediated gene suppression,[100] will provide new approaches to reveal the role of DEG/ENaC channels in sensory transduction.

Acknowledgments. The authors would like to acknowledge our colleagues who over the years have made important contributions to portions of the work summarized in this review including Drs. Xiuying Ma, Vladislav Snitsarev, Zhiyong Tan, Rasna Sabharwal, Heather Drummond, J. Thomas Cunningham, Margaret Sullivan, George Hajduczok, Ram V. Sharma, Margaret Price, Christopher Benson, Peter Snyder, and Michael J. Welsh. We are grateful also to Dr. Peter Snyder and Dr. Christopher Benson for scientific review of the manuscript and to Shawn Averkamp and Cheryl Ridgeway for secretarial assistance.

This publication was made possible by grant HL14388 from the National Institutes of Health, a VA Merit Award to MWC from the Department of Veterans Affairs, and funds from the Heartland Affiliate of the American Heart Association. Its contents are solely the responsibility of the authors and do not necessarily represent the official views of the NIH, VA, or AHA.

References

1. Landgren, S., On the excitation mechanism of the carotid baroceptors, *Acta. Physiol. Scand.* **26**, 1–34 (1952).
2. Kirchheim, H.R., Systemic arterial baroreceptor reflexes, *Physiol. Rev.* **56**, 100–176 (1976).
3. Brown, A.M., Receptors under pressure: an update on baroreceptors, *Circ. Res.* **46**, 1–10 (1980).
4. Chapleau, M.W., Li, Z., Meyrelles, S.S., Ma, X., and Abboud, F.M., Mechanisms determining sensitivity of baroreceptor afferents in health and disease, *Annals NY Acad. Sci.* **940**, 1–19 (2001).
5. Kellenberger, S., and Schild, L., Epithelial sodium channel/degenerin family of ion channels: a variety of functions for a shared structure, *Physiol. Rev.* **82**, 735–767 (2002).
6. Benson, C.J., Xie, J., Wemmie, J.A., Price, M.P., Henss, J.M., Welsh, M.J., and Snyder, P.M., Heteromultimers of DEG/ENaC subunits from H^+-gated channels in mouse sensory neurons, *Proc. Natl. Acad. Sci. USA* **99**, 2338–2343 (2002).
7. Drummond, H.A., Meyrelles, S.S., Price, M.P., Adams, C.M., Welsh, M.J., and Abboud, F.M., A molecular component of the arterial baroreceptor mechanotransducer, *Neuron* **21**, 1435–1441 (1998).

8. Price, M.P., Lewin, G.R., Mcllwrath, S.L., Cheng, C., Xie, J., Heppenstall, P.A., Stucky, C.L., Mannsfeldt, A.G., Brennan, T.J., Drummond, H.A., Qiao, J., Benson, C.J., Tarr, D.E., Hrstka, R., Yang, B., Williamson, R.A., and Welsh, M.J., The mammalian sodium channel BNC1 is required for normal touch sensation, *Nature* **407**, 1007–1011 (2000).
9. Hamill, O.P., and Martinac, B., Molecular basis of mechanotransduction in living cells *Physiol. Rev.* **81**, 685–740 (2001).
10. Welsh, M.J., Price, M.P., and Xie, J., Biochemical basis of touch perception: mechanosensory function of degenerin/epithelial Na^+ channels, *J. Biol. Chem.* **277**, 2369–2372 (2002).
11. Price, M.P., Mcllwrath, S.L., Xie, J., Cheng, C., Qiao, J., Tarr, D.E., Sluka, K.A., Brennan, T.J., Lewin, G.R., and Welsh, M.J., The DRASIC cation channel contributes to the detection of cutaneous touch and acid stimuli in mice, *Neuron* **32**, 1071–1083 (2001).
12. Snitsarev, V., Whiteis, C.A., Abboud, F., and Chapleau, M.W., Mechanosensory transduction of vagal and baroreceptor afferents revealed by study of isolated nodose neurons in culture, *Auton. Neurosci.* **98**, 59–63 (2002).
13. Tavernarakis, N., and Driscoll, M., Molecular modeling of mechanotransduction in the nematode *Caenorhabditis elegans*, *Annu. Rev. Physiol.* **59**, 659–689 (1997).
14. Lu, Y., Whiteis, C.A., Tan, Z-Y., Chapleau, M.W., and Abboud, F.M., Differential expression of acid-sensing ion channel (ASIC) subunits in rat carotid body (abstract), *FASEB J.* **20** (Pt. 2), A1230 (2006).
15. Guharay, F., and Sachs, F., Stretch-activated single ion channel currents in tissue-cultured embryonic chick skeletal-muscle, *J. Physiol. (London)* **352**, 685–701 (1984).
16. Hajduczok, G., Chapleau, M.W., Ferlic, R.J., Mao, H.Z., and Abboud, F.M., Gadolinium inhibits mechanoelectrical transduction in rabbit carotid baroreceptors, *J. Clin. Invest.* **94**, 2392–2396 (1994).
17. Driscoll, M., and Chalfie, M., The mec-4 gene is a member of a family of *Caenorhabditis elegans* genes that can mutate to induce neuronal degradation, *Nature* **349**, 588–593 (1991).
18. Ma, X.Y., Abboud, F.M., and Chapleau, M.W., Analysis of afferent, central, and efferent components of the baroreceptor reflex in mice, *Am. J. Physiol. Regul. Integr. Comp. Physiol.* **283**, R1033–R1040 (2002).
19. Cunningham, J.T., Wachtel, R.E., and Abboud, F.M., Mechanical stimulation of neurites generates as inward current in putative aortic baroceptor neurons in vitro, *Brain Res.* **757**, 149–154 (1997).
20. Sullivan, M.J., Sharma, R.V., Wachtel, R.E., Chapleau, M.W., Waite, L.J., Bhalla, R.C., and Abboud, F.M., Non-voltage-gated Ca^{2+} influx through mechanosensitive ion channels in aortic baroceptor neurons, *Circ. Res.* **80**, 861–867 (1997).
21. Sharma, R.V., Chapleau, M.W., Hajduczok, G., Wachtel, R.E., Waite, L.J., Bhalla, R.C., and Abboud, F.M., Mechanical stimulation increases intracellular calcium concentration in nodose sensory neurons, *Neurosci.* **66**, 433–441 (1995).
22. Kraske, S., Cunningham, J.H., Hajduczok, G., Chapleau, M.K., Abboud, F.M., and Wachtel, R.E., Mechanosensitive ion channels in putative aortic baroreceptor neurons, *Am. J. Physiol. Heart. Circ. Physiol.* **275**, H1497–H1501 (1998).
23. Cunningham, J.T., Wachtel, R.E., and Abboud, F.M., Mechanosensitive currents in putative aortic baroreceptor neurons in vitro, *J. Neurophysiol.* **73**, 2094–2098 (1995).

24. Chalfie, M., and Sulston, J., Developmental genetics of the mechanosensory neurons of *Caenorhabditis elegans*, *Dev. Biol.* **82**, 358–370 (1981).
25. Canessa, C.M., Horisberger, J.D., and Rossier, B.C., Epithelial sodium channel related to proteins involved in neurodegeneration, *Nature* **361**, 467–470 (1993).
26. Lingueglia, E., Voilley, N., Waldmann, R., Lazdunski, M., and Barbry, P., Expression cloning of an epithelial amiloride-sensitive Na^+ channel: a new channel type with homologies to *Caenorhabditis elegans* degenerins, *FEBS Lett.* **318**, 95–99 (1993).
27. Canessa, C.M., Schild, L., Buell, G., Thorens, B., Gautschi, I., Horisberger, J.D., and Rossier, B.C., Amiloride-sensitive epithelial Na^+ channel is made of three homologous subunits, *Nature* **367**, 463–466 (1994).
28. McDonald, F.J., Price, M.P., Snyder, P.M., and Welsh, M.J., Cloning and expression of the beta- and gamma-subunits of the human epithelial sodium channel, *Am. J. Physiol.* **268**, C1157–1163 (1995).
29. Garbers, D.L., and Dubois, S.K., The molecular basis of hypertension, *Annu. Rev. Biochem.* **68**, 127–155 (1999).
30. Pathak, B.G., Shaughnessy, J.D., Meneton, P., Greeb, J., Shull, G.E., Jenkins, N.A., and Copeland, N.G., Mouse chromosomal location of three epithelial sodium channel subunit genes and an apical sodium chloride cotransporter gene, *Genomics* **33**, 124–127 (1996).
31. de la Rosa, D.A., Canessa, C.M., Fyfe, G.K., and Zhang, P., Structure and regulation of amiloride-sensitive sodium channels, *Annu. Rev. Physiol.* **62**, 573–594 (2000).
32. Chen, C.C., England, S., Akopian, A.N., and Wood, J.N., A sensory neuron-specific, proton-gated ion channel, *Proc. Natl. Acad. Sci.* **95**, 10240–10245 (1998).
33. Waldmann, R., Champigny, G., Voilley, N., Lauritzen, I., and Lazdunski, M., The mammalian degenerin MDEG, an amiloride-sensitive cation channel activated by mutations causing neurodegeneration in *Caenorhadbditis elegans*, *J. Biol. Chem.* **271**, 10433–10436 (1996).
34. Adams, C.M., Price, M.P., Snyder, P.M, and Welsh, M.J., Tetraethylammonium block of BNC1 channel, *Biophys. J.* 76, 1377–1383 (1999).
35. Bassilana, F., Champigny, G., Waldmann, R., deWeille, J.R., Heurteaux, C., and Lazdunski, M., The acid-sensitive ionic channel subunit ASIC and the mammalian degenerin MDEG form a heteromultimeric H^+-gated Na^+ channel with novel properties, *J. Biol. Chem.* **272**, 28819–28822 (1997).
36. Price, M.P., Snyder, P.M, and Welsh, M.J., Cloning and expression of a novel human brain Na^+ channel, *J. Biol. Chem.* **271**, 7879–7882 (1996).
37. Garcia-Anoveros, J., Derfler, B., Neville-Golden, J., Hyman, B.T., and Corey, D.P, BNaC1 and BNaC2 constitute a new family of human neuronal sodium channels related to degenerins and epithelial sodium channels, *Proc. Natl. Acad. Sci. USA* **94**, 1459–1464 (1997).
38. Waldmann, R., Champigny, G., Bassilana, F., Heurteaux, C., and Lazdunski, M., A proton-gated cation channel involved in acid-sensing, *Nature* **386**, 173–177 (1997).
39. Waldmann, R., Bassilana, F., deWeille, J., Champigny, G., Heurteaux, C., and Lazdunski, M., Molecular cloning of a non-inactivating proton-gated Na^+ channel specific for sensory neurons, *J. Biol. Chem.* **272**, 20975–20978 (1997).
40. Lingueglia, E., deWeille, J.R., Bassilana, F., Heurteaux, C., Sakai, H., Waldmann, R., and Lazdunski, M., A modulatory subunit of acid-sensing ion channels in brain and dorsal root ganglion cells, *J. Biol. Chem.* **272**, 29778–29783 (1997).

41. Waldmann, R., and Lazdunski, M., H^+-gated cation channels: neuronal acid sensors in the ENaC/DEG family of ion channels, *Curr. Opin. Neurobiol.* **8**, 418–424 (1998).
42. Grunder, S., Geissler, H.S., Bassler, E.L., and Ruppersberg, J.P., A new member of acid-sensing ion channels from pituitary gland, *Neuroreport* **11**, 1607–1611 (2000).
43. Bassler, E.L., Ngo-Ann, T.J., Geisler, H.S., Ruppersberg, J.P., Grunder, S., Molecular and functional characterization of acid-sensing ion channel (ASIC)1b, *J. Biol. Chem.* **276**, 33782–33787 (2001).
44. Firsov, D., Gautschi, I., Merillat, A-M., Rossier, B.C., and Schild, L., The heterotetrameric architecture of the epithelial sodium channel (ENaC), *EMBO J.* **17**, 344–352 (1998).
45. Kosari, F., Sheng, S.H., Li, J.Q., Mak, D.O.D., Foskett, J.K., and Kleyman, T.R., Subunit stoichiometry of the epithelial sodium channel, *J. Biol. Chem.* **273**, 13469–13474 (1998).
46. Mano, I., and Driscoll, M., DEG/ENaC channels: a touchy superfamily that watches its salts, *Bioessays* **21**, 568–578 (1999).
47. Schild, L., Lu, Y., Gautschi, I., Schneeberger, E., Lifton, R.P., and Rossier, B.C., Identification of PY motif in the epithelial Na channel subunits as a target sequence for mutations causing channel activation found in Liddle syndrome, *EMBO J.* **15**, 2381–2387 (1996).
48. Snyder, P.M., Cheng, C., Prince, L.S., Rogers, J.C., and Welsh, M.J., Electrophysiological and biochemical evidence that DEG/ENaC cation channels are composed of nine subunits, *J. Biol. Chem.* **273**, 681–684 (1998).
49. Eskandari, S., Snyder, P.M., Kreman, M., Zampighi, G.A., Welsh, M.J., and Wright, E.M., Number of subunits comprising the epithelial sodium channel, *J. Biol. Chem.* **274**, 27281–27286 (1999).
50. Ernstrom, G.G., and Chalfie, M., Genetics of sensory mechanotransduction, *Ann. Rev. Genet.* **36**, 411–453 (2002).
51. Hughey, R.P., Bruns, J.B., Kinlough, C.L., Harkleroad, K.L., Tong, Q., Carattino, M.D., Johnson, J.P., Stockand, J.D., and Kleyman, T.R., Epithelial sodium channels are activated by furin-dependent proteolysis, *J. Biol. Chem.* **279**, 18111–18114 (2004).
52. Wang, N., Butler, J.P., and Ingber, D.E., Mechanotransduction across the cell surface and through the cytosleton, *Science* **260**, 1124–1127 (1993).
53. Hamill, O.P., and McBride, D.W., Jr., The cloning of a mechano-gated membrane ion channel, *Trends Neurosci.* **17**, 439–443 (1994).
54. Grady, R.M., Zhou, H., Cunningham, J.M., Henry, M.D., Campbell, K.P., and Sanes, J.R., Maturation and maintenance of the neuromuscular synapse: genetic evidence for roles of the dystrophin-glycoprotein complex, *Neuron* **25**, 279–293 (2000).
55. Fricke, B., Lints, R., Stewart, G., Drummond, H., Dodt, G., Driscoll, M., and von During, M., Epithelial Na^+ channels and stomatin are expressed in rat trigeminal mechanosensory neurons, *Cell Tissue Res.* **299**,327–334 (2000).
56. Mannsfeldt, A.G., Carroll, P., Stucky, C.L., and Lewin, G.R., Stomatin, a MEC-5 like protein, is expressed by mammalian sensory neurons, *Mol. Cell. Neurosci.* **13**, 391–404 (1999).
57. Leonard, A.S., Yermolaieva, O., Hruska-Hageman, A., Askwith, C.C., Price, M.P., Wemmie, J.A., and Welsh, M.J., cAMP-dependent protein kinase phosphorylation of the acid-sensing ion channel-1 regulates its binding to the protein interacting with C-kinase-1, *Proc. Natl. Acad. Sci. USA* **100**, 2029–2034 (2003).

58. Khalsa, P.S., Ge, W., Uddin, M.Z., and Hadjiargyrou, M., Integrin $\alpha 2\beta 1$ affects mechano-transduction in slowly and rapidly adapting cutaneous mechanoreceptors in rat hairy skin, *Neurosci.* **129**, 447–459 (2004).
59. Patel, A.J., Lazdunski, M., and Honore, D., Lipid and mechano-gated 2P domain K^+ channels, *Curr. Opin. Cell. Biol.* **13**, 422–427(2001).
60. Drummond, H.A., Welsh, M.J., and Abboud, F.M., ENaC subunits are molecular components of the arterial baroreceptor complex, *Ann. NY Acad. Sci.* **940**, 42–47 (2001).
61. de la Rosa, D.A., Zhang, P., Shao, D., White, F., and Canessa, C.M., Functional implications of the localization and activity of acid-sensitive channels in rat peripheral nervous system, *Proc. Natl. Acad. Sci. USA* **99**, 2326–2331 (2002).
62. Lu, Y., Whiteis, C.A., Benson, C.J., Chapleau, M.W., and Abboud, F.M., Expression and localization of acid-sensing ion channels in mouse nodose ganglia (abstract), *FASEB J.* **20** (Pt. 1), A775 (2006).
63. Deval, E., Baron, A., Lingueglia, E., Mazarguil, H., Zajac, J.M., Lazdunski, M., Effects of neuropeptide SF and related peptides on acid-sensing ion channel 3 and sensory neuron excitability. *Neuropharmacology* **44**, 662–671(2003).
64. Tan, Z-Y., Whiteis, C.A., Lu, Y., Chapleau, M.W., and Abboud, F.M., Role of BK and ASIC in the activation of glomus cells by extracellular acidosis and hypoxia (abstract), *FASEB J.* **20** (Pt. 2), A1230 (2006).
65. Babinski, K., Catarsi, S., Biagini, G., and Seguela, P., Mammalian ASIC2a and ASIC3 subunits coassemble into heteromultimeric proton-gated channels sensitive to Gd^{3+}, *J. Biol. Chem.* **275**, 28519–28525 (2000).
66. Xie, J.H., Price, M.P., Berger, A.L., and Welsh, M.J., DRASIC contributes to pH-gated currents in large dorsal root ganglion sensory neurons by forming heteromultimeric channels, *Neurophysiol.* **87**, 2835–2843 (2002).
67. Hesselager, M., Timmermann, D.B., and Ahring, P.K., pH dependency and desensitization kinetics of heterologously expressed combinations of acid-sensing ion channel subunits, *J. Biol. Chem.* **279**, 11006–11015 (2004).
68. Krishtal, O.A., and Pidoplichko, V.I., A receptor for protons in the nerve cell membrane, *Neurosci.* **5**, 2325–2327 (1980).
69. Krishtal, O.A., and Pidoplichko, V.I., A receptor for protons in the membrane of sensory neurons may participate in nociception, *Neurosci.* **6**, 2599–2601 (1981).
70. Bevan, S., and Yeats, J., Protons activate a cation conductance in a subpopulation of rat dorsal root ganglion neurons, *J. Physiol. (London)* **433**, 145–161 (1991).
71. Garcia-Anoveros, J., Samad, T.A., Woolf, C.J., and Corey, D.P., Transport and localization of the DEG/ENaC ion channel BNC1a to peripheral mechanosensory terminals of dorsal root ganglia neurons, *J. Neurosci.* **21**, 2678–2686 (2001).
72. Benson, C.J., Eckert, S.P., and McCleskey, E.W., Acid-evoked currents in cardiac sensory neurons: A possible mediator of myocardial ischemic sensation, *Circ. Res.* **84**, 921–928 (1999).
73. Sutherland, S.P., Benson, C.J., Adelman, J.P., and McCleskey, E.W., Acid-sensing ion channel 3 matches the acid-gated current in cardiac ischemia-sensing neurons, *Proc. Natl. Acad. Sci. USA* **98**, 711–716 (2001).
74. Meller, S.T., and Gebhart, G.F., A critical review of the afferent pathways and the potential chemical mediators involved in cardiac pain, *Neuroscience* **48**, 501–524 (1992).
75. Schultz, H.D., Cardiac vagal chemosensory afferents. Function in pathophysiological states, *Ann. NY Acad. Sci.* **940**, 59–73 (2001).

76. Snitsarev, V., Whiteis, C.A., Chapleau, M.W., and Abboud, F.M., Effect of prostacyclin analog on mechanosensitive vs. voltage-gated ion channels in nodose neurons (abstract), *Soc. Neurosc.* **27**, 1812 (2001).
77. Hummler, E., Barker, P., Gatzy, J., Beermann, F., Verdumo, C., Schmidt, A., Boucher, R., and Rossier, B.C., Early death due to defective neonatal lung liquid clearance in αENaC-deficient mice, *Nat. Genet.* **12**, 325–328 (1996).
78. Barker, P.M., Nguyen, M.S., Gatzy, J.T., Grubb, B., Norman, H., Hummler, E., Rossier, B., Boucher, R.C., and Koller, B., Role of γENaC subunit in lung liquid clearance and electrolyte balance in newborn mice, *J. Clin. Invest.* **102**, 1634–1640 (1998).
79. McDonald, F.J., Yang, B.L., Hrstka, R.F., Drummond, H.A., Tarr, D.E., McCray, P.B., Stokes, J.B., Welsh, M.J., and Williamson, R.A., Disruption of β subunit of the epithelial Na^+ channel in mice: hyper-kalemia and neonatal death associated with a pseudohypoaldosteronism phenotype, *Proc. Natl. Acad. Sci. USA* **96**, 1727–1731 (1999).
80. Ma, X.Y., Price, M.P., Drummond, H.A., Welsh, M.J., Chapleau, M.W., and Abboud, F.M., The DEG/ENaC ion channel family member BNC1 mediates mechanical transduction of arterial baroreceptor nerve activity in vivo (abstract), *FASEB J.* **15**, A1146 (Pt. 2) (2001).
81. Sabharwal, R., Stauss, H.M., Lazartigues, E., Whiteis, C.A., Davisson, R.L., Price, M.P., Welsh, M.J., Abboud, F.M., and Chapleau, M.W., Abnormalities in baroreflex sensitivity and autonomic control in conscious $ASIC2^{-/-}$ mice (abstract), *FASEB J.* **20** (Pt. 2) A1186 (2006).
82. Sun, W., Abboud, F.M., and Chapleau, M.W., Evaluation of baroreflex and chemoreflex by carotid artery occlusion in mice a method for phenotypic analysis of deletion of candidate sensory molecules (abstract), *Circulation*, **102**(18), II–700 (2000).
83. Sabharwal, R., Chapleau, M.W., Price, M.P., Welsh, M.J., and Abboud, F.M., Molecular mechanisms of baro- and chemoreceptor activation: evidence that ASIC1 and ASIC3 contribute to chemoreceptor activation (abstract), *Circulation* **112**(17), II–156 (2005).
84. Drew, L.J., Rohrer, D.K., Price, M.P., Blaver, K.E., Cockayne, D.A., Cesare, P., Wood, J.N., Acid-sensing ion channels ASIC2 and ASIC3 do not contribute to mechanically activated currents in mammalian sensory neurons, *J. Physiol. (London)* **556**, 691–710 (2004).
85. Roza, C., Puel, J.L., Kress, M., Baron, A., Diochot, S., Lazdunski, M., and Waldmann, R., Knockout of the ASIC2 channel in mice does not impair cutaneous mechanosensation, visceral mechanonociception, and hearing, *J. Physiol. (London)* **558**, 659–669 (2004).
86. Hildebrand, M.S., Silva, M.G., Klockars, T., Rose, E., Price, M., Smith, R.J.H., McGuirt, W.T., Christopoulos, H., Petit, C., and Dahl, H.H.M., Characterization of DRASIC in the mouse inner ear, *Hear. Res.* **190**, 149–160 (2004).
87. Peng, B.G., Ahmad, S., Chen, S.P., Chen, P., Price, M.P., and Lin, X., Acid-sensing ion channel 2 contributes a major component to acid-evoked excitatory responses in spiral ganglion neurons and plays a role in noise susceptibility of mice, *J. Neurosci.* **24**, 10167–10175 (2004).
88. Snitsarev, V., Whiteis, C.A., Chapleau, M.W., and Abboud, F.M., Neuronal prostacyclin is an autocrine regulator of arterial baroreceptor activity, *Hypertension* **4**, 540–546 (2005).

89. Li, Z., Lee, H.C., Bielefeldt, K., Chapleau, M.W., and Abboud, F.M., The prostacyclin analogue carbacyclin inhibits Ca^{2+}-activated K^{+} current in aortic baroreceptor neurones of rats, *J. Physiol.* **501**, 275–287 (1997).
90. Snitsarev, V., Sullivan, M.J., Whiteis, C.A., Chapleau, M.W., and Abboud, F.M., Molecular basis of mechanotransduction in baroreceptor neurons: roles of calcium-activated K current (I-KCa) and actin (abstract), *Circulation* **108**, IV-57 (2003).
91. Li, Z., Mao, H., Abboud, F.M., and Chapleau, M.W., Oxygen-derived free radicals contribute to baroreceptor dysfunction in atherosclerotic rabbits, *Circ. Res.* **79**, 802–811 (1996).
92. Li, Z., Chapleau, M.W., Bates, J.N., Bielefeld, K., Lee, H.C., and Abboud, F.M., Nitric oxide as an autocrine regulator of sodium currents in baroreceptor neurons, *Neuron* **20**, 1039–1049 (1998).
93. Snitsarev, V., Yermolaieva, O., Whiteis, C. A., Abboud, F. M., Heinemann, S. H., Hoshi, T., and Chapleau, M. W., Reactive oxygen species generated during action potential discharge mediate "activity-dependent resetting" of baroreceptor and vagal afferent neurons in culture (abstract), *Circulation* **106**, 66 (2002).
94. Francis, J., Weiss, R.M., Wei, S.G., Johnson, A.K., and Felder, R.B., Progression of heart failure after myocardial infarction in the rat, *Am. J. Physiol. Regul. Inter. Comp. Physiol.* **281**, R1734–1745 (2001).
95. Meyrelles, S.S., Mao, H.Z., Heistad, D.D., and Chapleau, M.W., Gene transfer to carotid sinus in vivo—a novel approach to investigation of baroreceptors, *Hypertension* **30** (Pt. 2), 708–713 (1997).
96. Kasparov, S., Teschemacher, A.G., Hwang, D.Y., Kim, K.S., Lonergan, T., and Paton, J.F.R., Viral vectors as tools for studies of central cardiovascular control, *Progr. Biophys. Mol. Biol.* **84**, 251–277 (2004).
97. Phillips, M.I., Gene therapy for hypertension: sense and antisense strategies, *Expert Opin. Biol. Therapy* **1**, 655–662 (2001).
98. Stewart, S.A., Dykxhoorn, D.M., Palliser, D., Mizuno, H., Yu, E.Y., An, D.S., Sabatini, D.M., Chen, I.S.Y., Hahn, W.C., Sharp, P.A., Weinberg, R.A., and Novina, C.D., Lentivirus-delivered stable gene silencing by RNAi in primary cells, *RNA-A Pub. RNA Soc.* **9**, 493–501 (2003).
99. Novina, C.D., and Sharp, P.A., The RNAi revolution, *Nature* **430**, 161–164 (2004).
100. Krug, U., Ganser, A., and Koeffler, H.P., Tumor suppressor genes in normal and malignant, *Oncogene* **21**, 3475–3495 (2002).

Part II
The TRP Family

4
TRP Channels as Molecular Sensors of Physical Stimuli in the Cardiovascular System

Roger G. O'Neil*

4.1. Introduction

The sensory systems of higher organisms utilize ion channels to transduce sensory stimuli into electrical signals. The sensory channels are either directly activated, such as observed for some mechanically sensitive channels (e.g., in touch), or indirectly activated by chemical components of a transduction pathways, such as observed for phototransduction and other pathways.[1,2] Independent of these specialized sensory cells, most cells in living organisms have the ability to sense and respond to alterations in local chemical and physical stimuli. In general, the molecular components of these cellular sensors and their transduction pathways are poorly understood.

In the past decade, new discoveries have begun to elucidate the identity of ion channels underlying "sensing" of certain chemical and physical stimuli. This includes the discovery of the DEG/ENaC superfamily, which may be a critical component of touch sensation in some organisms,[3,4] and the cyclic nucleotide-gated (CNG) channel family, which may be a component of odor sensation and other sensing processes.[2,5] Other channels may play an equally important role in sensing environmental changes. Indeed, the newly discovered transient receptor potential (TRP) family of channel proteins is beginning to emerge as an important component of numerous sensory processes. The first member of the channel family was identified in *Drosophila* as part of the phototransduction process where *Drosophila* mutants were discovered that displayed only transient receptor potentials (*trp*) when exposed to continuous light.[6,7]

The gene product was identified as a new type of cationic channel.[7] The subsequent identification of numerous TRP homologues in mammals that displayed nonselective cationic permeation properties gave rise to a new class of ion channel, the TRP protein family of ion channels.[1,8–11] While the function of many of the

*Department of Integrative Biology and Pharmacology, 6431 Fannin Street, Medical School, The University of Texas Health Science Center at Houston, Houston, TX 77030; roger.g.oneil@uth.tmc.edu

TRP channels is not known, a role in cellular sensing is beginning to emerge for some TRP family members.

This review is focused on the TRP channels as molecular sensors of physical stimuli. Other functions of TRP channels have been extensively reviewed by others.[1,8,12–14] Emphasis will be placed on the potential function of TRP channels in sensing changes in microenvironmental physical stimuli, with a special emphasis on cells of the cardiovascular system (see Table 4.1).

4.2. The TRP Superfamily

New members of the TRP channel superfamily continue to be discovered and now represent one of the largest families of ion channels with over 28 genes identified in the mammalian genome. The channels differ significantly from the voltage-activated calcium channels in structure and function. Individual TRP channels are either not voltage-sensitive or display weak voltage sensitivity. The channels are nonselective cation channels which, with few exceptions, are permeable to divalent cations such as calcium (see Refs. 8, 11). The calcium permeability is the basis of the channel function in controlling calcium influx into cells and, hence, in channel transduction and regulation of calcium signaling of many cells. The channels have a similar structure to the voltage-gated and CNG channels with six transmembrane domains (with the exception of the polycystin TRPP1), a pore-loop segment between transmembrane segments 5 and 6, and cytoplasmic N- and C- termini (Fig. 4.1) (see Refs. 1, 8, 10, 11).

Unlike the voltage-activated channels, transmembrane segment 4 contains few positively charged residues, which is consistent with a weak voltage sensitivity. Some subfamily members contain ankyrin repeat domains in their N-termini: 3-6 ankyrin repeat domains for TRPCs and TRPVs; 14 ankyrin repeats in TRPA1. A conserved TRP box (EWKFAR) is present in the C-termini domain of TRPCs and TRPMs. The channels are thought to assemble as tetramers to form functional channels, although an assessment of all TRP channels has not been completed. Further, functional TRP channels may require co-assembly with other TRP isoforms which may contribute to variations in activation and function of the channels (see Ref. 8).

The TRP channel subfamilies are classified by primary amino acid sequence and not by ion selectivity or ligand function. This largely relates to the varied properties and lack of specific ligands or inhibitors that are normally used to separate out activation pathways or physiological functions. Currently, the mammalian TRP superfamily is divided into six subfamilies, with only the first three being uniformly accepted.[8,15,16] The subfamilies are: TRPCs, the classical or canonical TRPs most closely related to the *Drosophila* TRP; TRPVs, vanilloid receptor-like; TRPMs, melastatin-like; TRPPs, polycystin-like; TRPA, ankyrin transmembrane protein-like (currently only one member, TRPA1 or ANKTM1); and TRPMLs, mucolipin-like. The last two subfamilies have not been widely studied and, hence, will not be discussed further in this chapter (see Ref. 8). In general, the TRP channels are

TABLE 4.1. The four dominant TRP channel subfamilies tissue distribution and properties

TRP name (prior names)	Tissue distribution*	Regulation/Activation**
TRPCs: Classical or canonical type		
TRPC1 (TRP1)	Widely expressed	GPCR-mediated, SOC?
TRPC2 (TRP2)	VNO, brain, heart, testis	GPCR-mediated
TRPC3 (TRP3)	Heart, muscle, placenta	GPCR-mediated, DAG, SOC?
TRPC4 (TRP4, CCE2)	Brain, testis, placenta, adrenal gland, kidney epithelial cells, endothelial cells	GPCR-mediated
TRPC5 (TRP5, CCE2)	Brain	GPCR-mediated
TRPC6	Brain, lung, muscle	GPCR-mediated, DAG, SOC?
TRPC7	Brain, heart, muscle, lung, eye	GPCR-mediated, DAG
TRPVs: Vanilloid receptor-like		
TRPV1 (VR1)	Brain, spinal cord, sensory neurons, bladder epithelial cells, skin	Capsaicin, anandamide, $T > 43\,^{\circ}C$, acidity
TRPV2 (VRL-1, GRC)	Brain, spinal cord, lung, spleen, sensory neurons, kidney epithelial cells, endothelial cells	$T > 52\,^{\circ}C$, hypotonicity, growth factors
TRPV3	Sensory neurons	$T > 30\,^{\circ}C$
TRPV4 (OTRPC4, TRP12, VR-OAC)	Kidney, trachea, skin, hypothalamus, epithelial cells, endothelial cells, VSM, heart	Hypotonicity, shear stress, $T > 25\,^{\circ}C$, phorbol esters, anandamide, arachodonic acid, epoxyeicosatrienoic acid
TRPV5 (ECaC1, CaT2)	Kidney, intestine, placenta, epithelial cells	Constitutive
TRPV6 (ECaC2, CaT1)	Intestine, placenta, kidney, prostate, salivary gland	Constitutive
TRPMs: Melastatin-like		
TRPM1 (Melastatin, LTRPC1)	Eye	ND
TRPM2 (TRPC7, LTRPC2)	Brain, placenta	ADP ribose, βNAD, H_2O_2?
TRPM3 (LTRPC3)	Kidney, tumor cells	Constitutive?, hypotonicity
TRPM4 (LTRPC4)	Widely expressed, including in kidney, heart, CNS, VSM	Stretch, $[Ca^{2+}]i$?
TRPM5 (Mtr1, LTRPC5)	VSM, taste receptor cells	GPCR-mediated
TRPM6 (ChaK 1, TRP-PLIK)	Intestine, kidney	Constitutive?
TRPM7	Kidney, heart, liver, spleen, lung, brain	Constitutive?, PIP_2, tyrosine phosphate
TRPM8 (Trp-p8, CMR1)	Sensory neurons, prostate carcinoma cells	$T < 25\,^{\circ}C$
TRPPs: Polycystin-like		
TRPP1 (PC1, PKD1)	Kidney, brain, heart, bone, muscle, VSM	ND
TRPP2 (PC2, PKD2)	Kidney, pancreas, liver lung, intestine, brain, thymus, reproductive organs, placenta, VSM, cardiomyocytes	Mechanical stress
TRPP3 (PKD2L1)	Kidney, VSM, heart	ND
TRPP5 (PKD2L2)	Kidney, VSM, heart	ND

*Regulation/Activation: DAG: diacylglycerol; GPCR: G-proein coupled receptor; PIP_2: phosphatidylinositol-4,5-bisphosphate; SOC: store-operated Ca^{2+} channel; ND: not determined
**Tissue Distribution: CNS: central nervous system; VNO: vomeronasal organ; VSM: vascular smooth muscle

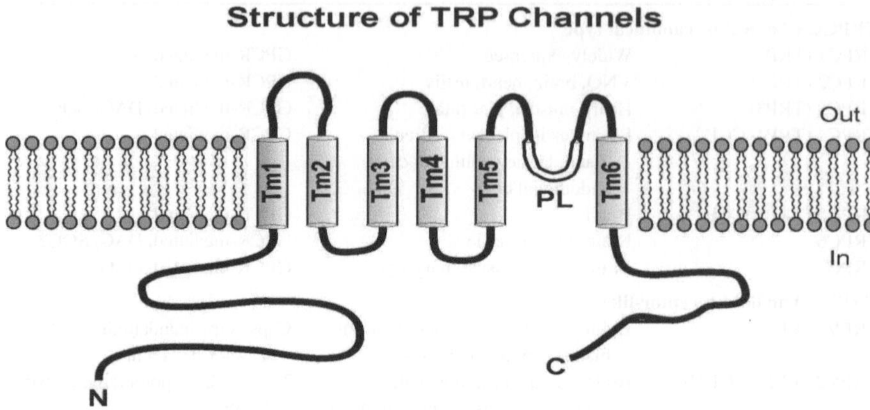

FIGURE 4.1. TRP channel topological model of channel structure. The typical TRP channel consists of six transmembrane domains (TM1–TM6), a pore loop (PL), and long N-terminus and C-terminus, both cytoplasmic. Some subfamily members contain ankyrin repeat domains (ARD) in their N-termini: 3-6 ankyrin repeat domains for TRPCs and TRPVs; 14 ankyrin repeats in TRPA1. A conserved TRP box (EWKFAR) is present in the C-termini domain of TRPCs and TRPMs. Unlike the voltage-activated channels, the TM4 transmembrane segment has few positively charged residues, accounting for the apparent weak voltage sensitivity of the channels.

widely expressed in a broad range of tissues, including both sensory and nonsensory cells, although some, such as TRPC2 and TRPV3, display a more limited distribution (see Table 4.1).

4.3. Overview of TRP Channel Function and Regulation

The physiological functions of the various TRP channels depend on the particular subfamily and specific channel. Early studies indicate that the functions are diverse, although the function of most TRP channels has remained elusive. A role in sensory biology was initially suspected by analogy to the *Drosophila* TRP channel which plays a central role in phototransduction of the fly eye.[1,9] While all the molecular details are not known, it has been clearly demonstrated that in *Drosophila*, activation of the TRP channels involves absorption of light by the G-protein-coupled rhodopsin which, in turn, activates phospholipase C to generate DAG and IP_3. The subsequent step has not been fully elucidated, but does not appear to involve IP_3 or release of calcium from stores as originally envisioned.

Recent evidence points to DAG, or its downstream polyunsaturated fatty acid (PUFA) metabolites (e.g., arachidonic acid, linolenic acid), as the ligand for activating the TRP channel. Other evidence points to a potential contributing role of decreased PIP_2 levels, following activation of PLC, which may relieve a block of the TRP channels. The mechanism of activation of the mammalian TRPC channels appears to have some analogy to the *Drosophila* TRP channel. All members of the family can be activated either via G-protein-coupled receptors and production of DAG, or by depletion of PIP_2 through activation of phospholipase C, as observed for *Drosophila* TRP, or, alternatively, by depletion of intracellular calcium stores (store-operated).[1,8,11] However, most mammalian TRP channels are not known to play a specific sensory role, although each could be part of a larger sensory transduction process. Indeed, TRPC2, expressed in the vomeronasal organ of rats and mice, appears to play such a key role in pheromone signaling because TRPC2-defiencient mice display abnormal mating behavior.[17] A specific role for TRPC channels in sensing physical stimuli has not yet been described.

The non-TRPC TRP channels may play a role in sensing physical and chemical stimuli.[8,11,12,18] Increasing evidence points to members of the TRPV subfamily in sensing both noxious and non-noxious stimuli ranging from temperature, to osmolality/stretch, to flow/shear stress and possible pressure. Likewise, some members of TRPM have a marked sensitivity to cold temperatures while some members of the TRPP subfamily appear to be sensitive to fluid flow. Hence, it is becoming apparent that many members of the TRP superfamily may act as sensors, or components of sensors, of physical stimuli. There is currently no evidence to suggest that the channels are activated directly by physical stimuli, such as observed for certain stretch-activated channels,[3] implying that the channels may be secondarily activated by molecular components of the sensory transduction pathway. Nonetheless, many of the TRP channels are widely expressed in sensory and nonsensory cells, including cells of the cardiovascular system and, hence, may play an important role in sensing physical stimuli and, in turn, modulating the function of the cardiovascular system. The potential role of these TRP channels in sensing physical stimuli, including in the cardiovascular system, are summarized below.

4.4. TRP Channels as Sensors of Physical Stimuli

4.4.1. Temperature

The ability to sense and respond to changes in temperature is critical to the survival of animals, especially warm-blooded animals such as mammals and birds. The molecular basis for sensing temperature may be closely associated with some of the vanilloid-receptor (TRPV) protein members. In 1997, the vanilloid receptor, or capsaicin receptor, was cloned and found to act as a nonselective cation channel.[19] Because the channel has significant sequence homology with the *Drosophila* TRP channel and with the subsequently identified mammalian TRPC channels, it was named as the founding member of the TRPV subfamily, TRPV1 (originally called VR1).[16] While the channel was found to be activated by vanilloid compounds, such as capsaicin (the hot ingredient of red peppers), it was also demonstrated to be activated by noxious temperatures above 43°C.[19] TRPV1-deficient mice had a reduced ability to sense painful thermal and chemical stimuli, but not mechanical stimuli, confirming that TRPV1 was sensitive to noxious temperature and chemical stimuli.[20] This provided the first evidence indicating the potential role of a mammalian TRP channel as a temperature sensor. The dominant expression of TRPV1 in small to medium-diameter neurons of the dorsal root, and trigeminal and nodose sensory ganglia, further supported the concept that TRPV1 is an integral component in temperature sensing.[12]

Other members of the TRPV subfamily, namely, TRPV2-4, were subsequently identified and shown to have high-temperature sensitivities for activation ($Q_{10} > 10$) and, therefore were candidates for temperature sensing. TRPV2 (previously called VRL-1) was sensitive to noxious temperatures above 52°C.[21] In contrast, TRPV3 and TRPV4 were sensitive to temperatures near the physiological range with TRPV3 being sensitive to activation in the 30–37°C range[22–24] and TRPV4 (previously called OTRPC4, TRP12, VR-OAC) in the 25–35°C range.[25,26]

The temperature sensitivity and threshold for activation of each channel may themselves be variable and regulated by other modulators of the channel such as observed for TRPV1 in the presence of the inflammatory mediator, bradykinin.[27] A variable sensitivity to stimuli may also be important for TRPV4 with temperature playing a critical role.[28] The sites of expression of TRPV channels may also support a temperature sensing role. TRPV2 is highly expressed in medium- to large-diameter fibers of thermosensory neurons, but is also present in the spinal cord, brain, spleen, and lung,[21] as well as in rat and mouse brain microvascular endothelial cells (Brown and O'Neil, unpublished data). TRPV3, on the other hand, was found to be expressed both in sensory ganglia and in temperature-sensitive tissues such as skin and tongue where it may play a role in sensing moderate temperatures.[22–24]

TRPV4 is expressed in some sensory neurons, but is expressed more prominently in the temperature-sensing areas of the anterior hypothalamus, as well as in numerous other tissues and cells such as kidney, trachea, liver, heart, lung, testis, brain, cochlea, and vascular endothelial cells.[26,29–31] While a temperature-sensing role of TRPV4 has yet to be clearly established, its recent identification in mouse skin keratinocytes,[32] which also express TRPV3, and subsequent demonstration of two temperature-induced currents consistent with TRPV3 and TRPV4 currents, provides compelling evidence of an integrated temperature sensing capability by the TRP channels in these cells.[33]

New studies have also revealed that certain TRP channels may also be sensitive to "cold" stimulation. Mammals have the ability to sense cold temperatures with the feeling of pain below 15°C.[14] A member of the TRPM subfamily, TRPM8 (previously called Trp-p8), was sensitive to cold stimulation with channel activation triggered by reduction of the temperature to the range of 25°C–8°C, or upon application of cooling agents, such as menthol.[34,35] Menthol appeared to act by increasing the temperature threshold, displaying a variable sensitivity with modulator agents, similar to that observed for TRPV1. In addition, like TRPV1, TRPM8 is expressed in a subset of small-diameter neurons in the trigeminal and dorsal root ganglia.[34,36] More recently, the sole member of the TRPA subfamily, TRPA1 (or ANKTM1), was reported to be activated by cold temperatures,[37] although this has been questioned by others.[38] It would appear, therefore, that TRPV1-4 and TRPM8 have sensitivities to detect temperatures between approximately 8°C and 50°C. These channels may play a central role in sensing and responding to temperature changes. Whether other undiscovered TRP channels or other ion channels are part of our sensory mechanisms for detecting temperature remains to be determined.[39]

Do TRP channels play a role in sensing temperature in the cardiovascular system? Temperature-sensing functions of the cardiovascular systems is not a known property of the system. However, two temperature-sensitive channels have been reported to be expressed in vascular endothelial cells, TRPV2 and TRPV4[40] (Brown and O'Neil, unpublished data). The expression of these two channels, particularly TRPV4, raises the possibility that modest, non-noxious temperatures could be sensed. This could play a role during periods of fever or local inflammation where modest increases in temperature occur. The potential activation, or enhanced activation, of the channel(s) may lead to an increased calcium influx into endothelial or smooth muscle cells which may be part of the inflammatory responses underlying the increased endothelial permeability or enhanced NO production and vasodilatation.[41] Future studies of the temperature sensitivity of TRP channels in these tissues will be required to assess the potential role of TRP channels in sensing cellular temperatures.

4.4.2. Osmotic Stress/Stretch

The TRPV4 channel was initially cloned based on homology with the *C. elegans* OSM-9 protein.[29–31] OSM-9 is part of the osmosensing and mechanosensing (touch) functions in worms.[42,43] While OSM-9 function as an active channel has not been demonstrated, TRPV4 expression in heterologous expression systems has been shown to function as an osmo-sensitive, calcium-permeable, cationic channel.[29–31] The subsequent observation that at room temperature the channel could be activated by the non-PKC-dependent phorbol ester, 4α-PDD,[40] opened the door to further demonstrations that at physiological temperatures (37°C), the channel was also sensitized to activation by PKC-dependent phorbol esters, demonstrating dual activation by PKC-dependent and PKC-independent pathways.[28,44] Hence, swelling-induced activation and translocation of PKC shown for some cells[45,46] may represent a potential TRPV4 modulating pathway.

Further, arachidonic acid, a phospholipase A2 metabolite, was found to generate cytochrome P450 epoxygenase metabolites that could activate the TRPV4 channel.[47] This may be part of the pathway for osmotic-induced activation of the channel because hypoosmotic swelling can activate PLA2.[47] The channel is widely viewed as an osmo-sensitive channel, although it is unlikely to be directly activated by hypotonic swelling as heretofore noted. Activation of the channel does rely on cell swelling because it was demonstrated in patch clamp studies that exposure of the cells to an internal and external hypoosmotic media did not activate the channel unless the cells were induced to swell.[48] Hence, stretch-induced activation of the channel is inferred and would be consistent with activation of the channel by fluid flow/shear stress (see below). However, in separate whole-cell patch studies done at room temperature, application of pressure to the back of the patch pipette to expand cell volume, failed to activate the channel.[30] It should be noted that because the later studies were done at room temperature and not at 37°C, application of simple "stretch" may fail to activate the channel as was observed for shear stress at lower temperatures (see below, Ref. 28). Hence, there may be a temperature-dependent conformational change or subunit association required for stretch-activation that is not apparent at room temperature, but is at physiological temperature. It remains for future studies to fully delineate the stretch-induced activation properties.

In addition to TRPV4, initial evidence demonstrates that TRPV2 may also be sensitive to hypoosmotic stress.[49] The channel is activated by hypoosmotic swelling implying a component of "stretch"-induced activation. Again, as for TRPV4, a direct demonstration of activation of TRPV2 by application of stretch to the cells has not been performed.

Recent studies have shown numerous TRP channels are expressed in smooth muscle cells, including TRPC, TRPV, and TRPM members (see Ref. 50). While the TRPC channel would appear to be involved in receptor-mediated regulation of calcium influx and membrane potential (see above), the mechanical sensitivity of the other channels may underlie myogenic regulation or stretch-induced vasoconstriction of small arteries. Specifically, both TRPM4 and TRPM5 have been shown to be expressed in cerebral artery smooth muscle cells and may play a role in myogenic constriction. Earley and coworkers[51] demonstrated that "stretching" cerebral arteries by application of pressure-induced calcium influx lead is depolarization of the membrane potential and constriction of the vessel.[51] Antisense oligodeoxynucleotide-induced downregulation of TRPM4 expression, but not TRPM5 expression, attenuated the stretch-induced vasoconstriction and abolished the depolarization of membrane potential. KCl-induced depolarization still induced a modest vasoconstriction, as expected, due to the separate activation of L-type calcium channels which carry a much smaller calcium influx. Jia and coworkers[52] have also shown expression of TRPV4 in airway smooth muscle cells that display constriction upon hypotonic swelling, supposedly reflecting stretch-induce activation of TRPV4 leading to calcium influx. Hence, accumulating evidence points to a potential role of selected members of both the TRPM and TPRV subfamilies in sensing mechanical stretch in vascular and other smooth muscle cells.

Vascular endothelial cells are also known to sense and respond to mechanical stimuli such as shear stress (see below) and/or stretch.[41,53,54] Expression of most TRPC channels in endothelial cells is now well documented.[54] Again, current evidence points to regulation of the TRPC channels by receptor-mediated mechanism coupled to activation of phospholipase C (see above). Further, a potential role of TRPC4-mediated calcium influx was found to be critical for a thrombin-induced increase in pulmonary endothelial permeability because the effects were abolished in TRPC4-deficient mice.[55]

Are other TRP channel members expressed in endothelial cells that could underlie stretch-induced calcium influx? Expression of TRPV4 has been identified in aortic endothelial cells[54] while both TRPV4 and TRPV2 have been shown to be expressed in rat and mouse brain microvascular endothelial cells (Brown and O'Neil, unpublished data). Hypotonic exposure of aortic endothelial cells in culture leads to activation of a calcium influx component consistent with activation of TRPV4[40] and/or TRPV2. However, direct demonstration of activation of TRPV4/TRPV2 or other TRP channels by cell stretch or increased pressure (leading to stretch) has not, as yet, been performed. Hence, the potential role of these channels in sensing and responding to mechanical stretch remains to be fully elucidated (see below).

4.4.3. Fluid Flow/Shear Stress

If certain TRP channels are sensitive to osmotic swelling and stretch, it may be that these channels are also sensitive to other forms of mechanical stress such as fluid flow or shear stress. Endothelial cells have long been known to display fluid flow/shear stress-dependent influx of calcium. We have shown that for TRPV4 in heterologous expression systems, increased fluid flow/shear stress activates the channel leading to calcium influx.[28] The activation is negligible at room temperature, but readily apparent at physiological temperatures (37°C), potentially indicating that a critical temperature-dependent conformational change or subunit association is needed for mechanical activation. More recently, we have shown that endogenous TRPV4 expressed in M-1 renal epithelial cells is also activated by fluid flow/shear stress when studied at 37°C.[56] TRPV4 knockdown with siRNA gene silencing techniques abolishes the shear stress-activated calcium influx, demonstrating that TRPV4 activation is sensitive to fluid flow. These studies provide strong evidence that TRPV4 is mechanically regulated in the endogenous setting, at least in epithelial cells, and that different forms of mechanical stress, including hypoosmotic swelling and fluid flow, can activate the channel. Whether other TRPV channels are also sensitive to shear stress has not been addressed.

Aortic[40] and rat and mouse brain microvessel endothelial cells (Brown and O'Neil, unpublished data) express TRPV4. Hence, it is likely that alterations in blood flow through capillaries and arteries can be sensed by TRPV4 channels and, possibly, other TRP channels yet to be identified. Such a flow-sensitive channel would account for the flow-induced increase in calcium influx that is widely reported for endothelial cells.[41] The increased calcium influx may, in part, underlie the flow-induced generation of endothelium-produced nitric oxide that plays a critical role in inducing relaxation of the vascular smooth muscle cells and vasodilatation.[41]

It has recently been shown that certain TRPP subfamily members may also be sensitive to fluid flow by mechanical bending of primary cilia in epithelial cells. TRPP1 and TRPP2 have been shown to localize to primary cilia in some renal tubular epithelial cells.[57,58] The primary cilia of renal epithelial cells are known to be sensitive to bending as occurs with fluid flow over the cell surface.[59] While TRPP1 is not known to form a functional channel, TRPP1 in association with TRPP2 does form a calcium-permeable channel.[57,60] Bending of the cilia leads to calcium influx into the cell, while applying a blocking antibody against TRPP2 abolishes the response.[57] Because other TRP channels are also expressed in the ciliated cells, it will be necessary for future studies to fully delineate the role of all channels in flow-induced calcium signaling before the basis of the flow-induced calcium signaling can be fully understood. While both TRPP1 and TRPP2 are widely expressed in various cells in the body,[61] including heart tissue and vascular smooth muscle cells (see Table 4.1), both nonciliated cells, a potential role for these channels in sensing flow seems unlikely, but this remains to be assessed.

4.4.4. Pressure

The electrical activity of baroreceptors in the aortic arch and carotid sinus are highly sensitive to alterations in mean arterial pressure. The molecular basis of the arterial baroreceptor as a pressure/mechanosensing and transduction pathway is poorly understood, but is thought to involve mechanosensitive channels, such as a DEG/ENaC (degenerin/epithelial Na^+ channel) complex;[62] also see Ref. 63. However, in an elegant series of studies by Abboud and coworkers[64,65] the effect of negative pressure in patch clamp studies (suction) applied to the cell surface of baroreceptor neurons from nodose ganglia of rats was shown to induce a nonselective cation channel. The channel was blocked by Gd^{3+}, but not by inhibitors of L-type, voltage-activated calcium channels. Alternatively, in calcium imaging studies, application of defined fluid pressures to the cell surface was shown to induce cell deformation and, hence, mechanical stimulation, leading to activation of calcium influx. The pressure-induced mechanical stimulation resulted in defined increases in calcium influx that were directly related to stimulus intensity and, again, were shown to be separate from L-type calcium channels. That is, the induced calcium influx was voltage-independent. Further, the induced calcium influx was inhibited by Gd^{3+}, a characteristic of many mechanosensitive ion channels and many TRP channels. This mechanically induced calcium influx has properties consistent with TRP mechanosensitive channels such as TRPV4 and TRPM4 (see above) which display cationic selectivity, activation by mechanical stimuli, and inhibition by Gd^{3+}.

While it is not currently known if such TRP channels, or other TRP channels, are expressed in baroreceptor neurons, it seems highly probably that such channels may underlie the mechanosensitive calcium influx in these cells. Other mechanosensitive channel complexes, such as the DEG/ENaC complex, may also play a role in sensing mechanical/pressure stimulation by modulating cell membrane potential and electrical excitability of the cells, but are not likely the basis of the calcium influx. Hence, it will be interesting to determine what role TRP channels may play, if any, in baroreceptor sensing and transduction of changes in blood pressure.

4.5. Conclusions

TRP channels represent a superfamily of nonselective, calcium-permeable, cation channels that are widely expressed in sensory and nonsenory cells. While a role for TRPC channels in receptor-mediated regulation of calcium influx is becoming apparent, increasing evidence points to a potential role of other TRP subfamilies members as sensors of physical stimuli. Selected members of the TRPV, TRPM, and TRPP subfamilies display a sensitivity to temperature and mechanical stimulation consistent with the channels being sensors, or components of sensory transduction pathways, depending on the particular tissue and site of expression (Fig. 4.2).

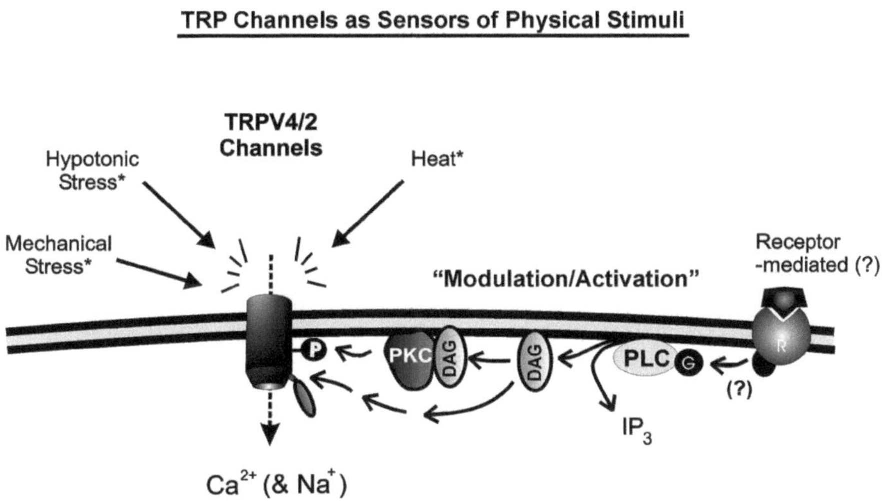

FIGURE 4.2. Molecular schema showing potential converging pathways of regulation of TRP channels by physical stimuli using TRPV4/TRPV2 as a general model. The TRPV4 and TRPV2 channels are permeable to calcium (Ca^{2+}) and monovalent cations such as sodium (Na^{+}). The channels appear to be regulated by at least three/four pathways/mechanisms: (1) mechanical /hypotonic stress, potentially via a phospholipase A2-mediated pathway,[47,49] (2) temperature, via an unknown pathway,[21,25,26] (3) DAG-induced, PKC-independent pathway,[28,40] and (4) DAG-induced, PKC-dependent pathway.[28,44] The PKC-dependent component may served to modulate or potentiate the channel to activation by other stimuli (see Ref. 28) and, hence, may act as a "modulatory" pathway for other stimuli such as observed for TRPV1.[68] R: hormone/ligand receptor; G: G-protein; PLC: phospholipase Cβ; DAG: diacylglycerol; PKC: protein kinase C; IP_3: inositol trisphosphate; P: potential phosphorylation site (?).

Some of the channels, such as TRPV1 and TRPV4, are sensitive to a wide range of physical and chemical stimuli where the channel may act as a site of integration of microenvironmental stimuli.[66,67] Many of these channels, including TRPV2, TRPV4, and TRPM4, are expressed within the cardiovascular system, including in endothelial cells and vascular smooth muscle cells, where they would appear to act as sensors of stretch, fluid flow, and local temperature. The importance of the various TRP channels in cardiovascular function is just beginning to emerge, setting the stage for discovery of many new and exciting functions and properties of this novel superfamily of ion channels.

References

1. Hardie, R.C., 2003, Regulation of TRP channels via lipid second messengers, *Annu. Rev. Physiol.* **65**:735–759.
2. Burns, M.E., and Baylor, D.A., 2001, Activation, deactivation, and adaptation in vertebrate photoreceptor cells, *Annu. Rev. Neurosci.* **24**:779–805.
3. Sukharev, S., and Corey, D.P., 2004, Mechanosensitive channels: multiplicity of families and gating paradigms, *Sci. STKE.* **2004** (219):re4.
4. Tavernarakis, N., and Driscoll, M., 2001, Mechanotransduction in *Caenorhabditis elegans*: the role of DEG/ENaC ion channels, *Cell Biochem. Biophys.* **35**:1–18.
5. Finn, J.T., Grunwald, M.E., and Yau, K.W., 1996, Cyclic nucleotide-gated ion channels: an extended family with diverse functions, *Annu. Rev. Physiol.* **58**:395–426.
6. Montell, C., and Rubin, G.M., 1989, Molecular characterization of the *Drosophila* trp locus: a putative integral membrane protein required for phototransduction, *Neuron* **2**:1313–1323.
7. Hardie R.C., and Minke B., 1992, The trp gene is essential for a light-activated Ca^{2+} channel in *Drosophila* photoreceptors, *Neuron* **8**:643–651.
8. Clapham, D.E., 2003, TRP channels as cellular sensors, *Nature* **426**:517–524.
9. Minke, B., and Cook, B., 2002, TRP channel proteins and signal transduction, *Physiol Rev.* **82**:429–472.
10. Montell, C., Birnbaumer, L., and Flockerzi, V., 2002a, The TRP channels, a remarkably functional family, *Cell* **108**:595–598
11. Voets, T., and Nilius, B., 2003, TRPs make sense, *J. Membr. Biol.*. **192**:1–8.
12. Benham, C.D., Gunthorpe, M.J., and Davis, J.B., 2003, TRPV channels as temperature sensors, *Cell Calcium* **33**:479–487.
13. Moran, M.M., Xu, H., and Clapham, D.E., 2004, TRP ion channels in the nervous system, *Curr. Opin. Neurobiol.* **14**:362–369.
14. Tominaga, M., and Caterina, M.J., 2004, Thermosensation and pain, *J. Neurobiol.* **61**:3–12.
15. Corey, D.P., 2003 New TRP channels in hearing and mechanosensation, *Neuron* **39**:585–588.
16. Montell, C., Birnbaumer, L., Flockerzi, V., Bindels, R.J., Bruford, E.A., Caterina, M.J., Clapham, D., Harteneck, C., Heller, S., Julius, D., Kojima, I., Mori, Y., Penner, R., Prawitt, D., Scharenberg, A.M., Schultz, G., Shimizu, S., and Zhu, M.X., 2002b, A unified nomenclature for the superfamily of TRP cation channels, *Mol. Cell* **9**:229–231.

17. Lucas, P., Ukhanov, K., Leinders-Zufall, T., and Zufall, F., 2003, A diacylglycerol-gated cation channel in vomeronasal neuron dendrites is impaired in TRPC2 mutant mice: mechanism of pheromone transduction, *Neuron* **40**:551–561.
18. Nauli, S.M., and Zhou, J., 2004, Polycystins and mechanosensation in renal and nodal cilia, *Bioessays* **26**:844–856.
19. Caterina, M.J., Schumacher, M.A., Tominaga, M., Rosen, T.A., Levine, J.D., and Julius, D., 1997, The capsaicin receptor: a heat-activated ion channel in the pain pathway, *Nature* **389**:816–824.
20. Caterina, M.J., Leffler, A., Malmberg, A.B., Martin, W.J., Trafton, J., Petersen-Zeitz, K.R., Koltzenburg, M., Basbaum, A.I., and Julius, D, 2000, Impaired nociception and pain sensation in mice lacking the capsaicin receptor, *Science* **288**:306–313.
21. Caterina, M.J., Rosen, T.A., Tominaga, M., Brake, A.J., and Julius, D., 1999, A capsaicin-receptor homologue with a high threshold for noxious heat, *Nature* **398**:436–441.
22. Peier, A.M., Reeve, A.J., Andersson, D.A., Moqrich, A., Earley, T.J., Hergarden, A.C., Story, G.M., Colley, S., Hogenesch, J.B., McIntyre, P., Bevan, S., and Patapoutian, A., 2002b, A heat-sensitive TRP channel expressed in keratinocytes, *Science* **296**:2046–2049.
23. Smith, G.D., Gunthorpe, M.J., Kelsell, R.E., Hayes, P.D., Reilly, P., Facer, P., Wright, J.E., Jerman, J.C., Walhin, J.-P., Ool, L., Egerton, J., Charles, K.J., Smart, D., Randall, A.D., Anand, P., and Davis, J.B., 2002, TRPV3 is a temperature-sensitive vanilloid receptor-like protein, *Nature* **418**:186–190.
24. Xu, H., Ramsey, I.S., Kotecha, S.A., Moran, M.M., Chong, J.A., Lawson, D., Ge, P., Lilly, J., Silos-Santiago, I., Xie, Y., DiStefano, P.S., Curtis, R., and Clapham, D.E., 2002, TRPV3 is a calcium-permeable temperature-sensitive cation channel, *Nature* **418**:181–186.
25. Güler, A.D., Lee, H., Iida, T, Shimizu, I., Tominaga, M., and Caterina, M., 2002, Heat-evoked activation of the ion channel, TRPV4, *J. Neurosci.* **22**: 6408–6414.
26. Watanabe, H., Vriens, J, Suh, S.H., Benham, C.D., Droogmans, G., and Nilius, B., 2002b, Heat-evoked activation of TRPV4 channels in a HEK293 cell expression system and in native mouse aorta endothelial cells, *J. Biol. Chem.* **277**:47044–47051.
27. Chuang, H.H., Prescott, E.D., Kong, H., Shields, S., Jordt, S.E., Basbaum, A.I., Chao, M.V., and Julius, D., 2001, Bradykinin and nerve growth factor release the capsaicin receptor from PtdIns(4,5)P2-mediated inhibition, *Nature* **411**:957–962.
28. Goa, X., Wu, L., and O'Neil, R.G., 2003, Temperature-modulated diversity of TRPV4 channel gating: activation by physical stresses and phorbol ester derivatives through protein kinase C-dependent and independent pathways, *J. Biol. Chem.* **278**:27129–27137.
29. Liedtke, W., Choe, Y., Marti-Renom, M.A., Bell, A.M., Denis, C.S., Sali, A., Hudspeth, A.J., Friedman, J.M., and Heller, S., 2000, Vanilloid receptor-related osmotically activated channel (VR-OAC), a candidate vertebrate osmoreceptor, *Cell* **103**:525–535.
30. Strotmann, R., Harteneck, C., Nunnenmacher, K., Schultz, G., and Plant, T., 2000, OTRPC4, a nonselective cation channel that confers sensitivity to extracellular osmolarity, *Nature Cell Biol.* **2**:695–702.
31. Wissenbach, U., Bödding, M., Freichel, M., and Flockerzi, V., 2000, Trp12, a novel Trp related protein from kidney, *FEBS Letters* **485**: 127–134.
32. Chung, M.K., Lee, H., Caterina, M.J., 2003, Warm temperatures activate TRPV4 in mouse 308 keratinocytes, *J. Biol. Chem.* **278**:32037–32046.

33. Chung, M.K., Lee, H., Mizuno, A., Suzuki, M., and Caterina, M.J., 2004, TRPV3 and TRPV4 mediate warmth-evoked currents in primary mouse keratinocytes, *J. Biol. Chem.* **279**:21569–21575.
34. McKemy, D.D., Neuhausser, anf W.M., Julius, D, 2002, Identification of a cold receptor reveals a general role for TRP channels in thermosensation, *Nature* **416**:52–58.
35. Peier, A.M., Moqrich, A., Hergarden, A.C., Reeve, A.J., Andersson, D.A., Story, G.M., Earley, T.J., Dragoni, I., McIntyre, P., Bevan, S., and Patapoutian, A., 2002a, A TRP channel that senses cold stimuli and menthol, *Cell* **108**:705–715.
36. Nealen, M.L., Gold, M.S., Thut, P.D., and Caterina, M.J., 2003, TRPM8 mRNA is expressed in a subset of cold-responsive trigeminal neurons from rat *J. Neurophysiol.* **90**:515–520.
37. Story, G.M., Peier, A.M., Reeve, A.J., Eid, S.R., Mosbacher, J., Hricik, T.R., Earley, T.J., Hergarden, A.C., Andersson, D.A., Hwang, S.W., McIntyre, P., Jegla, T., Bevan, S., and Patapoutian, A., 2003, ANKTM1, a TRP-like channel expressed in nociceptive neurons, is activated by cold temperatures, *Cell* **112**:819–829.
38. Jordt, S.E., Bautista, D.M., Chuang, H.H., McKemy, D.D., Zygmunt, P.M., Hogestatt, E.D., Meng, I.D., and Julius, D., 2004, Mustard oils and cannabinoids excite sensory nerve fibres through the TRP channel ANKTM1, *Nature* **427**:260–265.
39. Viana, F., de la Pena, E., and Belmonte, C., 2002, Specificity of cold thermotransduction is determined by differential ionic channel expression, *Nat. Neurosci.* **5**:254–260.
40. Watanabe, H., Davis, J.B., Smart, D., Jerman J.C., Smith, G.D., Hayes, P., Vriens, J., Cairns, W., Wissenbach, U., Prenen, J., Flockerzi, V., Droogmans, G., Benham, C.D., and Nilius, B., 2002a, Activation of TRPV4 channels (hVRL-2/mTRP12) by phorbol derivatives, *J. Biol. Chem.* **277**:13569–13577.
41. Boo, Y.C., and Jo, H., 2003, Flow-dependent regulation of endothelial nitric oxide synthase: role of protein kinases, *Am. J. Physiol. Cell Physiol.* **285**:C499–C508.
42. Colbert, H.A., Smith, T.L., and Bargmann, C.I., 1997, OSM-9, a novel protein with structural similarity to channels, is required for olfaction, mechanosessation, and ofactory adaptation in *Caenorhabditis elegans*, *J. Neurosci.* **17**(21): 8259–8269.
43. Tobin, D., Madsen, D., Khan-Kirby, A., Peckol, E., Moulder, G., Barstead, R., Maricq, A., and Bargmann, C., 2002, Combinatorial expression of TRPV channel proteins defines their sensory functions and subcellular localization in *C. elegans* neurons, *Neuron* **35**: 307–318.
44. Xu, F., Satoh, E., and Iijima, T., 2003, Protein kinase C-mediated Ca2+ entry in HEK 293 cells transiently expressing human TRPV4, *Br. J. Pharmacol.* **140**:413–421.
45. Liu, X., Zhang, M.I.N., Peterson, L.B., and O'Neil, R.G, 2003, Osmomechanical stress selectively regulates translocation of protein kinase C isoforms, *FEBS Letters* **538**:101–106.
46. O'Neil, R.G., and Leng, L., 1997, Osmo-mechanically sensitive phosphatidylinositol signaling regulates a Ca^{2+} influx channel in renal epithelial cells, *Am. J. Physiol.* **273**:F120–128.
47. Vriens, J., Watanabe, H., Janssens, A., Droogmans, G., Voets, T., and Nilius, B., 2004, Cell swelling, heat, and chemical agonists use distinct pathways for the activation of the cation channel TRPV4, *Proc. Natl. Acad. Sci. USA* **101**:396–401.
48. Nilius, B., Prenen, J., Wissenbach, U., Bodding, M., and Droogmans, G., 2001, Differential activation of the volume-sensitive cation channel TRP12 (OTRPC4) and volume-regulated anion currents in HEK-293 cells, *Pflügers Arch.* **443**:227–233.

49. Muraki, K., Iwata, Y., Katanosaka, Y., Ito, T., Ohya, S., Shigekawa, M., and Imaizumi, Y., 2003, TRPV2 is a component of osmotically sensitive cation channels in murine aortic myocytes, *Circ. Res.* **93**:829–838.
50. Beech, D.J., Muraki, K., and Flemming, R., 2004, Nonselective cationic channels of smooth muscle and the mammalian homologues of Drosophila TRP, *J. Physiol.* **559**(Pt 3): 685–706.
51. Earley, S., Waldron, B.J., and Brayden, J.E., 2004, Critical role for transient receptor potential channel TRPM4 in myogenic constriction of cerebral arteries, *Circ. Res.*, **95**: 922–929.
52. Jia, Y., Wang, X., Varty, L., Rizzo, C.A., Yang, R., Correll, C.C., Phelps, P.T., Egan, R.W., and Hey, J.A., 2004, Functional TRPV4 channels are expressed in human airway smooth muscle cells, *Am. J. Physiol Lung Cell Mol Physiol.* **287**:L272–L278.
53. Ishida, T., Takahashi, M., Corson, M.A., and Berk, B.C., 1997, Fluid shear stress-mediated signal transduction: how do endothelial cells transduce mechanical force into biological responses? *Ann. NY Acad. Sci.* **811**:12–23
54. Nilius, B., Droogmans, G., Wondergem, R., 2003, Transient receptor potential channels in endothelium: solving the calcium entry puzzle? *Endothelium* **10**:5–15.
55. Tiruppathi, C., Freichel, M., Vogel, S.M., Paria, B.C., Mehta, D., Flockerzi, V., and Malik, A.B., 2002, Impairment of store-operated Ca2+ entry in TRPC4$^{-/-}$ mice interferes with increase in lung microvascular permeability, *Circ. Res.* **91**:70–76.
56. O'Neil, R.G., L. Wu, and X. Gao, 2003, Knockdown of the TRPV4 calcium-permeable channel in mouse M-1 cells and a heterologous expression system by small interfering RNA (siRNA), *J. Am. Soc. Nephrol.* **14**:82A
57. Nauli, S.M., Alenghat, F.J., Luo, Y., Williams, E., Vassilev, P., Li, X., Elia, A.E., Lu, W., Brown, E.M., Quinn, S.J., Ingber, D.E., and Zhou, J., 2003, Polycystins 1 and 2 mediate mechanosensation in the primary cilium of kidney cells, *Nat. Genet.* **33**:129–137.
58. Yoder, B.K., Hou, X., and Guay-Woodford, L.M., 2002, The polycystic kidney disease proteins, polycystin-1, polycystin-2, polaris, and cystin, are co-localized in renal cilia, *J. Am. Soc. Nephrol.* **13**:2508–2516.
59. Praetorius, H.A., and Spring, K.R., 2001, Bending the MDCK cell primary cilium increases intracellular calcium, *J. Membr. Biol.* **184**:71–79.
60. Hanaoka, K., Qian, F., Boletta, A., Bhunia, A.K., Piontek, K., Tsiokas, L., Sukhatme, V.P., Guggino, W.B., and Germino, G.G., 2000, Co-assembly of polycystin-1 and -2 produces unique cation-permeable currents, *Nature* **408**:990–994.
61. Igarashi, P., and Somlo, S., 2002, Genetics and pathogenesis of polycystic kidney disease, *J. Am. Soc. Nephrol.* **13**:2384–2398.
62. Drummond, H.A., Welsh, M.J., and Abboud, F.M., 2001, ENaC subunits are molecular components of the arterial baroreceptor complex, *Ann. NY Acad. Sci.* **940**:42–47.
63. Drummond, H.A., Gebremedhin, D., and Harder, D.R., 2004, Degenerin/epithelial Na^+ channel proteins. Components of a vascular mechanosensor, *Hypertension*, **44**: 643–648.
64. Kraske, S., Cunningham, J.T., Hajduczok, G., Chapleau, M.W., Abboud, F.M., and Wachtel, R.E., 1998, Mechanosensitive ion channels in putative aortic baroreceptor neurons *Am. J. Physiol.* **275**:H1497–H1501.
65. Sullivan, M.J., Sharma, R.V., Wachtel, R.E., Chapleau, M.W., Waite, L.J., Bhalla, R.C., and Abboud, F.M., 1997, Non-voltage-gated $Ca2^+$ influx through mechanosensitive ion channels in aortic baroreceptor neurons, *Circ. Res.* **80**:861–867.

66. Gunthorpe, M.J., Benham, C.D., Randall, A., Davis, J.B., 2002, The diversity in the vanilloid (TRPV) receptor family of ion channels, *Trends Pharmacol. Sci.* **23**: 183–191.
67. O'Neil, R.G., and Brown, R.C., 2003, The vanilloid receptor family of calcium-permeable channels: molecular integrators of microenvironmental stimuli, *News Physiol. Sci.* **18**:226–231.
68. Bhave, G., Hu, H.J., Glauner, K.S., Zhu, W., Wang, H., Brasier, D.J., Oxford, G.S., and Gereau, R.W., 4th., 2003, Protein kinase C phosphorylation sensitizes but does not activate the capsaicin receptor transient receptor potential vanilloid 1 (TRPV1), *Proc. Natl. Acad. Sci USA.* **100**:12480–12485.

5
TRPV1 in Central Cardiovascular Control

Discerning the C-Fiber Afferent Pathway

Michael C. Andresen*, Mark W. Doyle†, Timothy W. Bailey, and Young-Ho Jin

5.1. Introduction

Progress in understanding the central nervous system (CNS) mechanisms regulating cardiovascular function has long been linked to the neurobiology of cranial primary sensory afferents. Activation of visceral afferents with chemical substances provided seminal evidence that particular afferents even within a single organ (e.g., the heart) or sensory modality (e.g., mechanoreceptors) could have fundamentally different characteristics and evoke unique reflex outcomes. In cardiorespiratory afferent studies, early practitioners deployed a range of sometimes rather exotic exogenous compounds to probe the discharge properties of afferent nerves as well as to evoke reflex responses. These chemicals ranged from neurotransmitters, peptides, prostanoids, cytokines, phenylbiguanide, and veratridine to nicotine.[1–4] Thus, the pharmacology of primary visceral afferents is intimately interwoven into the fabric of CNS processing and the physiology of autonomic reflexes.

Few substances have enjoyed the longevity or the wide utility of another agent, capsaicin (CAP). Many decades before the cloning and mapping of the vanilloid receptor TRPV1, CAP defined the link of CAP to pain sensation to activation of a particular class of afferents. Early investigations confirmed what every consumer of piquant peppers understands—CAP activates primary nociceptive afferents that convey the painful, burning sensation of CAP application. These CAP-responsive sensory neurons have slowly conducting axons (C-type and thinly myelinated Aδ-type) and cell bodies in the dorsal root ganglia (DRG).[5] Similar strategies identified CAP-sensitive afferents within the viscera and an association with cardiovascular regulation.[2] Many of these visceral afferents are cardiovascular mechanoreceptors,[6] and their reflexes utilize vagal afferent pathways.[3] Such visceral afferents constitute a unique cranial

*Department of Physiology and Pharmacology, Oregon Health and Science University, Portland, Oregon USA 97239-3098. andresen@ohsu.edu
†Current address: Dept. of Biology, George Fox University, Newberg, Oregon, USA 97132-2697.

pathway to the CNS because they enter the brain, not through the spinal cord (e.g., [7]), but rather through cranial ganglia (e.g. nodose, NG) to directly enter the brainstem. Vagal cranial afferents course directly to neurons within the nucleus tractus solitarius (NTS) to form their first CNS synapses. Thus, NTS is truly the "gateway" through which visceral sensory information must pass to initiate a multitude of reflexes controlling autonomic and homeostatic organ regulation.[8,9]

Within the brainstem, early neurochemical probing identified multiple potential neurotransmitters using immunocytochemistry and microinjection within NTS.[10] These findings often paralleled neurotransmission in the superficial lamina of the spinal cord where visceral and somatic spinal afferents are processed and implicated glutamate, GABA and substance P but many additional neuropeptides were present within NTS.[11,12] For example, in the caudal NTS, glutamate agonist microinjection evoked reflex decreases in heart rate and blood pressure from the subnuclei associated with cardiovascular regulation.[13] These same regions were abundant in substance P immunoreactivity and introduction of substance P into these regions mimicked the baroreceptor reflex.[14,15] Within this context, CAP injected into NTS similarly decreased blood pressure and heart rate consistent with cardiovascular afferent activation. Thus, CAP has long been generally associated with sensory activation and together was part of the evidence supporting this region of NTS as the site of the primary afferent synapses subserving the baroreceptor reflex.[14,15]

With the cloning of TRPV1 in 1997,[16] focus converged on a specific molecular target for CAP and the vanilloid field exploded in new developments. This work has offered new insights to the large body of early observations as well as posing new mysteries about TRPV1 mechanisms of action and their functional significance. The mRNA for TRPV1 is localized to spinal region of the DRG—a region long associated with CAP and pain. TRPV1 and CAP sensitivity are thus selectively expressed in subsets of spinal sensory neurons, typically those with un- or lightly myelinated axons.[17–19] Native sensory neurons as well as heterologous expression systems demonstrated that TRPV1 acts as an ion channel with broad selectivity for cations but has a substantial preference for calcium over sodium ions (3:1). The most obvious difference in nodose neurons is in the expression of sodium channels sensitive to TTX[20,21]—a myelinated/unmyelinated difference that is broadly reminiscent of spinal sensory DRG neurons.[22,23] Interestingly, although the broad comparison between somatic afferent neurons and cranial visceral afferent neurons indicates general similarities, it is already clear that important functional differences exist. These critical details of ion channel expression may be responsible for interesting functional differences of these cranial afferents compared to their spinal cousins. CAP selectively binds to TRPV1 and triggers a large cationic flux that depolarizes these cells. A frequently overlooked observation of the initial report noted that TRPV1 mRNA also localized to NG, a group of cranial visceral sensory neurons not associated with nociception or pain pathways.[16] As with DRG neurons, CAP-sensitivity of NG neurons was widely reported some decades before the TRPV1 cloning.[18,24]

5.2. Cranial TRPV1: The Context Broadens

Despite this strong association of CAP with pain transduction, evidence since the early 1950s indicated that CAP was part of a varied group of chemical agents that activate afferents arising from visceral organs including the heart, lungs, and blood vessels. Fiber recordings included afferent neurons with axons within the vagus, i.e. cranial visceral afferents. Thus, for example, atrial mechanoreceptors with C-fiber conduction velocities are excited by CAP and this maneuver elicited vigorous reflex changes in blood pressure and heart rate.[1] Gastrointestinal vagal afferent neurons share a similar neurobiology of CAP and C-fiber sensory neurons.[25] Collectively, the most resonant similarity with the DRG literature is the association of CAP sensitivity with slow conduction velocities of the afferent axons recorded from NG neurons. Efferent vagal axons conducting in C-type range are not similarly affected by CAP.[26] Thus, conduction velocity or more specifically most C-type afferent axons appear to be sensitive to CAP irrespective of the location of their cell bodies or the destination of their central terminations (spinal cord/brain stem).

5.3. TRPV1: Primary Afferent Entree to Brainstem Pathways

The arterial baroreceptor represents an interesting special case in the realm of sensory neurobiology. In the rat, aortic baroreceptor axons are concentrated in a single, thin nerve trunk called the aortic depressor nerve (ADN). The ADN courses parallel to but rarely comingles with the vagal trunk of the tenth cranial nerve on its way to the NG. This anatomical segregation unique to the rat and rabbit means that the ADN contains only afferent baroreceptor axons. Most analogous nerves in other species are contaminated by large numbers of the chemoreceptor afferents common in such nerves as the carotid sinus nerve even in the rat. The usual functional evidence of chemoreceptor activity is manifested in reflexes but is lacking for ADN[27–30] including the conscious rat.[31] No respiratory or pressor responses are found when the rat ADN is activated unlike the carotid sinus nerve.[27] The vagus, like many peripheral nerves, contains a mixture of afferent axons from various target organs and sensory modalities plus efferent axons with varying destinations. The ADN is a rare case of a purely afferent sensory nerve trunk.

The electroneurogram of the ADN (Fig. 5.1) reveals two discrete populations of baroreceptor axons, one conducting in the Aδ and another in the C-fiber range (conducting at <0.5 m/sec).[32] Cross-sections of the ADN reveal that most (65–80%) axons in the rat ADN are unmyelinated.[33,34] The peripheral structure of the sensory endings remains relatively simple even as it arborizes within the adventitial layer of the aortic arch.[35,36] A-fiber and C-fiber baroreceptor axons are intimately intertwined even at the distal arbors.[35] Myelinated and unmyelinated primary cranial afferent neurons express substantially different patterns of receptors and ion channels that imbue each neuron class with distinct action potentials, spike adaptation, and frequency limits.[21,37] The sensory mechanotransduction mechanism for

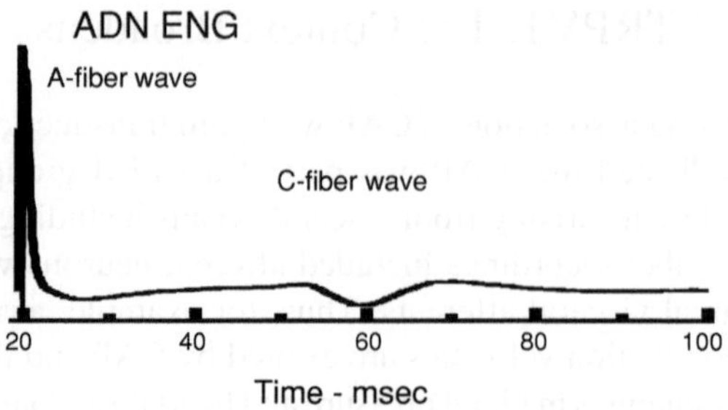

FIGURE 5.1. Electroneurogram of the rat aortic depressor nerve (ADN). A single supramaximal electrical shock elicits an early, fast-conducted (10–12 m/sec), A-fiber wave and a late arriving, slowly conducted (<0.5 m/sec) C-fiber wave. These two waves of compound action potentials reveal the underlying two distinct populations of baroreceptor axons. Periaxonal exposure to capsaicin between the stimulating and recording electrodes completely eliminates the conducted C-wave (not shown), evidence for a homogeneous expression of TRPV1 in ADN baroreceptor C-fiber peripheral axons.

baroreceptors appears directly related to stretching of the arterial wall with an ionic mechanism that is poorly understood but is likely to be broadly similar to other mechanoreceptors.[38,39] Unlike many C-fiber mechanoreceptors within the cardiovascular system that have been termed chemically sensitive, arterial baroreceptors appear to be relatively resistant to chemical activation.[1]

Myelinated axons have much lower thresholds for electrical activation than unmyelinated peripheral axons. Progressive increases in the intensity of electrical shocks to the ADN elicit at the lowest levels only an A-fiber volley and then as stimulus intensity is increased a C-fiber compound action potential is evoked that progressively increases in size as more axons are recruited to fire (Fig. 5.1). This differential sensitivity to stimulus intensity allows separate activation of A- and C-fiber baroreceptors and has been used to great advantage in studying the frequency response characteristics of the arterial baroreflex arising from A- and C-fiber baroreceptor pathways.[32,40,41] In this way, low intensity stimulation of ADN activates A-fiber baroreceptors without C-fiber co-activation. Similar electrical stimulation protocols suggest that A-selective and A+C baroreceptor activation evoke distinctly different stimulus frequency response relations for the baroreflex control of MAP or HR (Figure 5.2).[32,40,42,43]

Because baroreceptors had C-fiber axons, we decided to test whether periaxonal application of CAP would affect axonal conduction. CAP applied to isolated sections of the ADN selectively blocks the conduction along C-type axons.[32] This periaxonal CAP block eliminated the ADN C-wave but not the conducted A-wave. In blocking conduction then, CAP blocks C-type baroreceptor reflex responses but reveals the full response of the A-fiber baroreceptor reflex pathway at high frequencies and maximal activation.[32] These observations suggest clear functional differences in the performance characteristics of the respective pathways activated by A- and C-type baroreceptors. It is particularly interesting that these distinctions

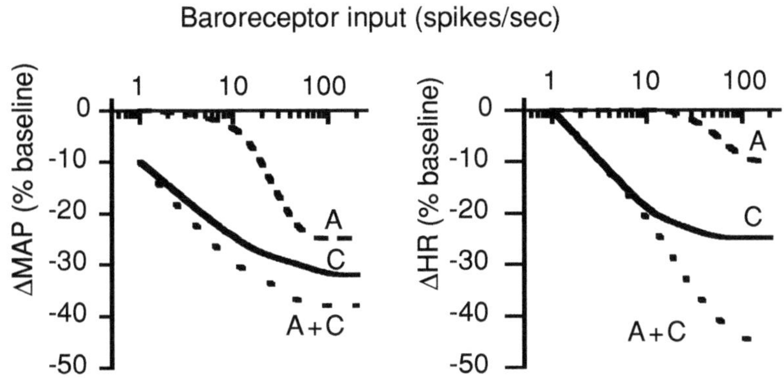

FIGURE 5.2. Baroreflex responses activated by electrical stimulation of the aortic depressor nerve (ADN) in anesthetized rats. Panels show changes in mean arterial blood pressure (MAP, left) and heart rate (HR, right) to 30 sec of ADN stimulation at frequencies ranging from 1 to 200 Hz. At low shock intensities, only A-fiber baroreceptors are activated (A, dashed line). At maximal shock intensities, all baroreceptors are activated (A+C, dotted line). At high shock intensities with steady anodal block, only C-fiber baroreceptor activity is conducted centrally (C, solid line). A-fiber baroreceptors are quite ineffective at activating baroreflex responses at frequencies of less than 10 Hz while C-fiber baroreceptors significantly alter cardiovascular regulation at much lower frequencies.

persist through the entire reflex pathway from afferent initiation to the final reflex outcome. The mechanisms responsible and the site(s) of action that give rise to the performance differences in these pathways are unknown. The possible presence of TRPV1 on unmyelinated visceral afferents and its absence on the myelinated afferents suggested a potential strategic opening to begin to delve into the CNS mechanisms of pathway differentiation.

5.3.1. Basics of the Baroreflex Pathway

Visceral afferents from NG distribute synapses across NTS neurons to some extent viscerotopically.[44,45] Cardiovascular afferents impinge on dorsomedial NTS, whereas respiratory afferent endings are found ventrally and ventrolaterally. There is clear overlap. NTS lies at the beginning portions of the overall basic schematic brainstem circuits of the baroreflex (Fig. 5.3). Important nuclei within the brainstem include NTS, the caudal ventrolateral medulla (CVLM), the rostral ventrolateral medulla, vagal dorsal motor nucleus (DMN), and nucleus ambiguus (NA).[46] The overall organization of the parasympathetic path is generally considered less convoluted than the sympathetic pathways with as few as two central neurons.[47] More detailed information about each step along these reflex pathways is imperative to assessing overall function and discerning the mechanisms responsible for the differences in reflex responses evoked by myelinated and unmyelinated baroreceptors.

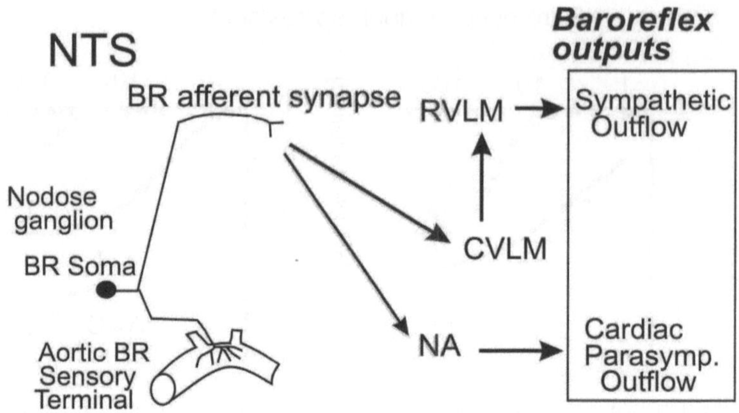

FIGURE 5.3. Schematic of visceral sensory afferents (e.g., baroreceptors, BR) to brain stem nuclei and autonomic pathways. BR fibers synapse within the nucleus tractus solitarius, NTS. This afferent initiated signal then branches to a sympathetic pathway and a cardiac parasympathetic pathway. The sympathetic pathway involves several nuclei within the brainstem before exiting to the spinal cord and then to cardiovascular targets via peripheral sympathetic postganglionic neurons near their targets. The central parasympathetic pathway exits the CNS directly to parasympathetic postganglionic neurons within the heart—an intrinsically shorter pathway requiring fewer than the sympathetic pathway.

5.4. NTS Synaptic Transmission

Our approach has focused on sensory transmission to NTS as a pragmatic starting point for studies of these brainstem pathways. Neurons that receive baroreceptor synapses are most concentrated within the dorsomedial portions of caudal NTS.[48–50] This region of NTS is a particularly important site within the overall reflex pathway because the primary afferent synaptic contact is an obligatory step of the baroreflex. Furthermore, considerable work points to NTS as the site of interactions between sensory and important supramedullary neurons[8] and evidence in particular implicates presynaptic afferent sites.[51–53] Thus, identifying those initial neurons with afferent contacts, that is, the second-order neurons, is a key step for these investigations.

The diffuse cellular organization of NTS presents an experimental challenge. The laminar neuron arrangement common in many CNS regions (e.g., the cortex)[54] is absent in NTS. Within NTS, the lack of functional segregation means that different functional classes (e.g., second- and higher-order neurons) are often found closely adjacent to each other and thus must be separated by particular strategies.[55] Another challenge to studies is the small size of the NTS neurons. *In vivo* intracellular recordings from these neurons tend to be quite short in duration and limit the nature and scope of the investigations that are possible.[56–59]

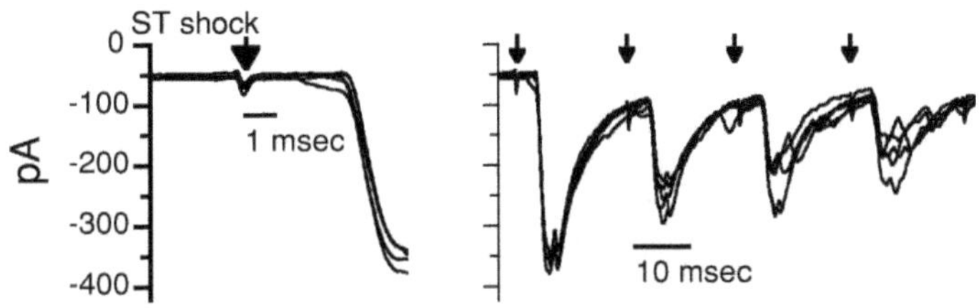

FIGURE 5.4. Reliable, nearly invariant excitatory postsynaptic currents (EPSCs) indicative of a monosynaptic response to solitary tract (ST) activation recorded in a second-order NTS neuron in brain stem slice. Arrows indicate ST shocks followed at a nearly constant delay (synaptic latency) of several milliseconds. The standard deviation of that latency is the jitter. Low-jitter (generally <200 μsec; note left panel), large amplitude (100s of pA) and frequency-dependent depression (note right panel) of ST EPSCs are characteristic of second-order NTS neurons including those anatomically identified with fluorescently dyed ADN terminals.

Concentrating on the medial sub-nucleus, we developed a horizontal brainstem slice to provide a stable and controlled preparation for studies of the cellular mechanisms of synaptic transmission.[60] These horizontal NTS slices are optimized to allow electrical activation of cranial visceral afferents during ST stimulation isolated from contamination by activation of local interneurons.[61,62] The afferent axons in these horizontal slices are visible as a longitudinal stripe of the solitary tract (ST) that is preserved in the same plane as the NTS neuron cell bodies. Second-order neurons are identified by their synaptic response characteristics to ST activation. Electrical stimuli delivered to the ST activate excitatory postsynaptic currents (EPSCs) with nearly invariant latencies (low jitter), and these EPSCs rarely fail to appear even at high frequencies of stimulation.[63] In most second-order neurons these EPSCs rely solely on non-NMDA receptors for fast glutamate synaptic transmission (Fig. 5.4). On the contrary, more complicated paths to higher-order neurons have highly variable latencies, and synaptic responses are often absent with rapid activation—signs indicative of a polysynaptic pathway. In addition, when lipophilic carbocyanine tracers including DiA are placed on the peripheral ADN trunk, the dye is transported centrally and fluorescently identifies neurons that receive baroreceptor synaptic boutons. These ADN dye labeled neurons have the same low-jitter EPSC responses characteristic of direct activation by ST synapses.[63] Thus, second-order neurons can be identified within brain slices of NTS by electrical and anatomical means.

One of the surprises from the initial slice studies of NTS second-order neurons was the unitary nature of ST synaptic responses. Unlike in the peripheral ADN stimulation studies, increasing the ST stimulus intensity evoked all-or-none responses rather than recruiting larger or more diverse responses as stimulus intensity increased. We were not able to electrically distinguish A- from C-fiber afferent axons despite their clear presence within these afferent nerves.

5.5. TRPV1 on Presynaptic ST Afferent Terminals

Given our experience with CAP on the peripheral ADN, we decided to test whether TRPV1 was present on the central synaptic endings with NTS and thus if CAP might be useful in identifying C-type afferent responses in brain stem slices. CAP applied to slices initially increased spontaneous synaptic activity.[64] At very low concentrations (1 nM), CAP facilitated ST-evoked EPSCs, but at high concentrations (100 nM), CAP evoked large inward currents and with continued exposure blocked ST synaptic transmission in a subgroup of second-order neurons in medial NTS—CAP-sensitive.[64] The CAP-evoked inward current was blocked by non-NMDA selective antagonists or by capsazepine. CAP affects both ADN dye-labeled baroreceptor as well as unlabeled second-order neurons.[48] Thus, CAP triggers an initial release of glutamatergic synaptic vesicles from the CAP-sensitive terminals of ST axons and during prolonged exposure vesicles are either depleted or the presynaptic process becomes inactivated by CAP. The expression of TRPV1 clearly defines two classes of ST afferents to NTS, CAP-sensitive and CAP-resistant, and these are consistent with C- and A-type peripheral cranial sensory afferent neurons (Fig. 5.5).

This CAP-based strategy of chemically defined second-order NTS neurons by the sensitivity of their afferent synaptic connections to CAP. This afferent differentiation of second-order NTS neurons, however, was accompanied by a second major surprise. These two classes of NTS neurons were clearly different in their

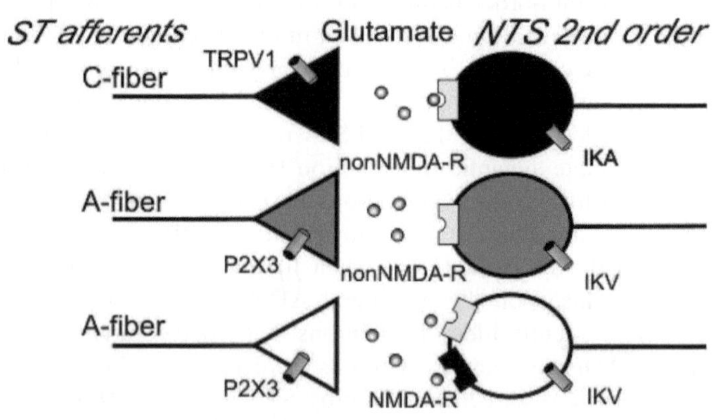

FIGURE 5.5. Schematic of TRPV1 identified afferent paths to NTS. Afferent glutamate-releasing terminals (left) are divided into C- and A-fiber classes based on the presence of TRPV1 and CAP-sensitivity. Note that additionally and alternatively, P2X3 expression and sensitivity to $\alpha\beta$ methylene ATP mark A-type terminals. Note that some A-type terminals activate neurons (right) with NMDA in addition to non-NMDA receptors postsynaptically. NTS neurons contacted by C-type afferents have prominent transient IKA sensitive to 4-aminopyridine, while all second-order NTS neurons have similar sustained potassium currents (IKV, TEA sensitive).

postsynaptic properties—their potassium channels. All second-order NTS neurons had similar TEA-sensitive potassium steady-state currents (IKV). However, CAP-sensitive second-order neurons had significantly greater transient, 4-AP sensitive potassium currents (IKA currents) than CAP-resistant second-order NTS neurons. Thus, the presence or absence of TRPV1 identifies two afferent pathways in NTS that differ in key properties at both their pre- as well as their postsynaptic makeup. Thus, these elements of the reflex pathways demonstrate that the pathways diverge in their central properties as early the first synapse. The potassium channel differences will have substantial impact on differences in the frequency tuning of these pathways to afferent discharge frequency. If the CAP-resistant pathway represents A-type afferents such as aortic baroreceptors, the relative absence of IKA will mean less spike frequency adaptation and perhaps greater transmission of the high-fidelity A-type baroreceptor activity. On the other hand in the C-type pathway at NTS, the prominent expression of postsynaptic IKA might act as a low-pass filter on afferent inputs in the CAP-sensitive (C-type) pathway. Thus, K^+ channels may be a key factor in performance differences of neuron discharge differentiating particular pathways at the earliest stages of NTS and in turn greatly influence the propagation of excitation through the full pathway to effector targets in the periphery.

5.6. Isolated NTS Neurons with Intact Boutons

CAP can be a problematic compound to control even in slice preparations. CAP is highly lipophilic, a property making it difficult to wash out and reverse its actions. A new phase of our TRPV1 work was greatly facilitated by a different and fundamentally reduced preparation—the mechanically dissociated NTS neuron. Neurons with intact synaptic boutons were first dissociated >15 yrs ago, a feat developed using NTS neurons.[65] Recent refinements suggest that dissociated neurons might help us better control experiments to address complex intercellular interactions.[66]

To dissociate NTS neurons within the medial sub-nucleus of NTS, we mechanically dissociated neurons from slices identical to those described above and directed the fire-polished pipette to our normal recording area. Using a micromanipulator under visual microscopic control, the pipette tip harvested NTS neurons from regions of medial NTS limited to where we find ADN dye labeling and generally record in slice work.[60,66] The resulting dissociated neurons were strikingly similar in dimensions and appearance to views of the soma and proximal processes observed in slices. Voltage clamp recordings from such dissociated neurons showed the presence of a spontaneous release of neurotransmitters onto all neurons as both spontaneous EPSCs (sEPSC) and as sIPSCs. NBQX blocked the fast kinetic sEPSCs and GABA bicuculline blocked the slower kinetic sIPSCs. Synaptic events also persisted in TTX and are termed miniature synaptic events (mEPSCs or mIPSCs) in which transmitter release is independent of action potentials.

5.7. TRPV1/P2X3 Identify C- and A-Type Afferents in NTS

CAP slice studies indicated that CAP specifically discriminated two major subclasses of cranial afferents to NTS. Third- or higher-order NTS neurons did not respond to CAP directly so that positive TRPV1 responses positively identify at least some neurons as second-order and not higher-order. Just as in slices, exposure of isolated NTS neurons to 100 nM CAP evoked a release of glutamatergic EPSCs by a presynaptic action (Fig. 5.6). CAP altered only the rate of glutamate release and not the amplitude of the EPSCs, indicating a selective presynaptic mechanism. Capsazepine potently antagonized CAP at 500 nM. GABA release was unaffected by CAP, demonstrating CAP specificity for glutamatergic terminals. One advantage of dissociated neurons is that both CAP and its antagonist reversed in seconds, a feat that would require some 30–45 min of washing in the slice. Responses could be repeated and readily combined with other more complicated manipulations with agonists given separately as well as in combination with antagonists, all within an individual neuron. As in slices, many isolated NTS neurons were unresponsive to CAP, yet these neurons had sEPSCs and thus glutamatergic terminals that might originate from A-type, myelinated afferent ST terminals.

With the precedent of the dorsal horn, we tested whether purinergic receptors might modulate glutamate release onto isolated NTS neurons and how this was related to TRPV1 distribution. Both ATP and the nonhydrolysable analog $\alpha\beta$-methylene ATP ($\alpha\beta$-m-ATP, 10 μM) triggered release of glutamate in dissociated NTS neurons. The pharmacological profile of the ATP responses suggested a P2X3 purinergic receptor subtype. Remarkably, testing single NTS neurons for both TRPV1 and P2X3 suggested that these receptors uniquely segregated to different presynaptic terminals and were found on different NTS neurons.[66]

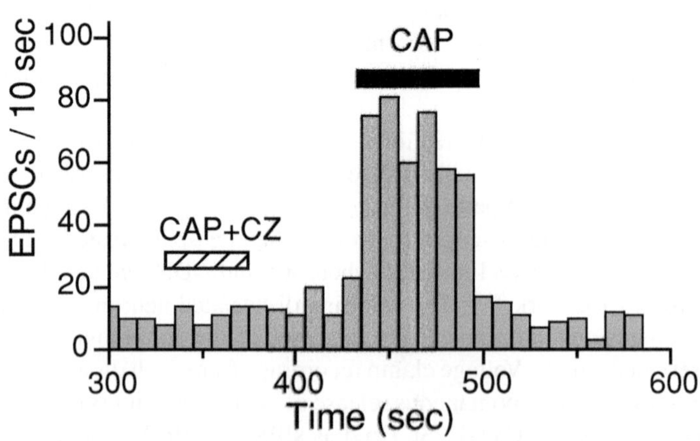

FIGURE 5.6. In NTS neurons dissociated from the dorsal medial sub-nucleus, retained glutamatergic synapses spontaneously release transmitter recorded as EPSCs. Application of capsaicin (CAP, 100 nM) rapidly and reversibly evokes an increased rate of glutamate release that is blocked by TRPV1 antagonist capsazepine (CZ, 500 nM).

Because these two receptors were distributed mutually exclusively to different terminals and different neurons, we turned to tests of NG neurons—the neuronal source of our presynaptic ST terminals—to test whether CAP and $\alpha\beta$-m-ATP responsiveness corresponded to C-conducting and A-conducting axons, respectively.[66] $\alpha\beta$-m-ATP evoked inward currents only in NG neurons with A-fiber conduction velocities and were CAP-resistant. Conversely, CAP evoked an inward current in CAP-sensitive NG neurons that had C-fiber conduction velocities and these neurons never responded to $\alpha\beta$-m-ATP. Such results confirm that CAP sensitivity is uniquely associated with cranial primary afferent neurons with C-fiber axons and not with Aδ fibers and provide a second, unique marker agonist profile for myelinated cranial afferents in medial NTS. Thus, the pharmacological profiles of afferent ST terminals indicates receptor segregation centrally and offers a new strategy to manipulate and identify A- and C-type pathways centrally in slices.

5.8. TRPV1 Permeability in NTS

Both TRPV1 and P2X3 are coupled to relatively nonselective cation channels.[67] Both channels have substantial Ca^{++} permeability. Activation of these channels in presynaptic afferent terminals then could contribute to the Ca^{++} that normally controls synaptic transmission. We tested for Ca^{++} permeation through presynaptic channels on glutamate release using dissociated NTS neurons.[66] Within single neurons, we measured with agonist release of glutamate during progressive, sequential blockade of voltage-gated Na^{+} channels using TTX and Ca^{++} channels with Cd^{++}. Even in the presence of block of presynaptic depolarization via Na^{+} and Ca^{++} channels, activation of either presynaptic TRPV1 or P2X3 receptors on their respective NTS neuron subtypes triggered substantial glutamate release. These two presynaptic targets may represent unique targets for intervention at the first step of brain stem pathways important to autonomic and homeostatic regulation.

5.9. TRPV1 Sorting the Reflex Path

This strategic localization of TRPV1 on cranial afferent presynaptic terminals may help to better define and investigate the pathway beyond NTS and its second-order neurons. Cardiac vagal preganglionic neurons in nucleus ambiguus (NA) are the final central station of the parasympathetic pathway that regulates heart function and the excitatory pathways that activate these critical neurons are poorly understood. Activation of ADN baroreceptors suggested a disproportionately powerful influence of C-type baroreceptors.[32,43] These cardiac NA neurons can be isolated for cellular electrophysiological study using retrograde dyes from the heart to identify selectively these neurons *in vitro*.[68,76,77–78] Stimulation of the vagus nerve in brain stem slices evoked EPSCs in identified cardiac vagal preganglionic neurons. Application of CAP to such brain slices substantially reduced the amplitude of synaptic responses to vagal activation. This CAP-senstive or C-type afferent pathway to cardiac vagal preganglionic neurons had a shorter latency and greater amplitude than the CAP-resistant, A-type pathway. Thus, the C-type pathway from cranial afferents to these cardiac vagal neurons may be intrinsically simpler and more powerful than the A-type pathway through the brain stem, although it is not clear what mechanisms are responsible for the relative differences in response magnitude. Such results are consistent with the full reflex loop heart rate responses from C-selective and A-selective ADN baroreceptor stimulation protocols.[43]

5.10. TRPV1: Sign Post at the Brain Stem Gateway

Combining these with other results, a model of the A- and C-fiber pathways is starting to take form. The C-fiber afferent terminates in TRPV1 containing terminals that release glutamate to activate postsynaptic non-NMDA receptors. Other differences likely are expressed at the junctions of these two major classes of presynaptic afferent terminals. The general anesthetic ketamine[69] blocks ST evoked EPSCs at lower concentrations in CAP-sensitive afferents than in CAP-resistant afferents (Fig. 5.5). Such evidence supports two subdivisions of A-fiber paths. A final factor in this pathway differentiation may reflect a general principle increasingly recognized in neurobiology.[70] Neuronal phenotype is partly intrinsic and established embryonically by gene expression (e.g., ion channels, myelination, etc.). Additionally differences are strongly influenced by network activity. Clearly, not all heterogeneities found in second-order NTS neurons are associated with this dichotomy of A- and C-pathways but may well contribute to afferent frequency-dependent action potential traffic that is characteristic of the sensory discharge patterns inherent in A/C afferents.[37,39,71,72] Thus, interconnections and their activity critically refine cellular heterogeneity.[70] The interplay with other neurons in a network context strongly selects for final features and ultimate performance characteristics of neurons and their pathways. Fundamentally, we do not fully understand how the complex gating and modulation of TRPV1[67,73,74] factors into the local environment within the brain stem and NTS is, for example, the intermittent activity of the cardiac or baroreceptor afferents an important aspect of driving these synaptic terminals in NTS? Clearly, many issues remain unresolved about the nature of the heterogeneity of these terminals as a part of the primary long-term objective of understanding the CNS information processing that sets the baroreflex and other cardiovascular control circuits.[75]

References

1. Coleridge, H.M., and Coleridge, J.C.G., Cardiovascular afferents involved in regulation of peripheral vessels, *Annu. Rev. Physiol.*, **42**, 413–427 (1980).
2. Coleridge, J.C.G., and Coleridge, H.M., Chemoreflex regulation of the heart. In: *Handbook of Physiology: The Cardiovascular System I*, American Physiological Society, Bethesda, MD, 1980, pp. 653–676.

3. Verberne, A.J., Saita, M., and Sartor, D.M., Chemical stimulation of vagal afferent neurons and sympathetic vasomotor tone, *Brain Res. Brain Res. Rev.*, **41**, 288–305 (2003).
4. Schultz, H.D., Cardiac vagal chemosensory afferents: Function in pathophysiological states, *Ann. NY. Acad. Sci.*, **940**, 59–73 (2001).
5. Foreman, R.D., Integration of viscerosomatic sensory input at the spinal level, *Prog. Brain Res.*, **122**, 209–221 (2000).
6. Coleridge, H.M., Coleridge, J.C.G., and Kidd, C., Role of the pulmonary arterial baroreceptors in the effects produced by capsaicin in the dog, *J. Physiol*, **170**, 272–285 (1964).
7. Zahner, M.R., Li, D.P., Chen, S.R., and Pan, H.L., Cardiac vanilloid receptor 1-expressing afferent nerves and their role in the cardiogenic sympathetic reflex in rats, *J. Physiol*, **551**, 515–523 (2003).
8. Andresen, M.C., and Kunze, D.L., Nucleus tractus solitarius: gateway to neural circulatory control, *Annu. Rev. Physiol.*, **56**, 93–116 (1994).
9. Browning, K.N., and Mendelowitz, D., Musings on the wanderer: What's new in our understanding of vago-vagal reflexes?: II. Integration of afferent signaling from the viscera by the nodose ganglia, *Am. J. Physiol. Gastrointest. Liver Physiol.*, **284**, G8–G14 (2003).
10. Van Giersbergen, P.L.M., Palkovits, M., and De Jong, W., Involvement of neurotransmitters in the nucleus tractus solitarii in cardiovascular regulation, *Physiol. Rev.*, **72**, 789–824 (1992).
11. Lawrence, A.J., and Jarrott, B., Neurochemical modulation of cardiovascular control in the nucleus tractus solitarius, *Prog. Neurobiol.*, **48**, 21–53 (1996).
12. Palkovits, M., Distribution of neuroactive substances in the dorsal vagal complex of the medulla oblongata, *Neurochem. Int.*, **7**, 213–219 (1985).
13. Talman, W.T., Granata, A.R., and Reis, D.J., Glutamatergic mechanisms in the nucleus tractus solitarius in blood pressure control, *Fed. Proc.*, **43**, 39–44 (1984).
14. Haeusler, G., and Osterwalder, R., Evidence suggesting a transmitter or neuromodulatory role for substance P at the first synapse of the baroreceptor reflex, *Naunyn-Schmiedeberg's Arch. Pharmacol.*, **314**, 111–121 (1980).
15. Lukovic, L., De Jong, W., and De Wied, D., Cardiovascular effects of substance P and capsaicin microinjected into the nucleus tractus solitarii of the rat, *Brain Res.*, **422**, 312–318 (1987).
16. Caterina, M.J., Schumacher, M.A., Tominaga, M., Rosen, T.A., Levine, J.D., and Julius, D., The capsaicin receptor: a heat-activated ion channel in the pain pathway, *Nature*, **389**, 816–824 (1997).
17. Caterina, M.J., and Julius, D., The vanilloid receptor: a molecular gateway to the pain pathway, *Annu. Rev. Neurosci.*, **24**, 487–517 (2001).
18. Szallasi, A., and Blumberg, P.M., Vanilloid (capsaicin) receptors and mechanisms, *Pharmacol. Rev.*, **51**, 159–212 (1999).
19. Stebbing, M.J., McLachlan, E.M., and Sah, P., Are there functional P2X receptors on cell bodies in intact dorsal root ganglia of rats?, *Neuroscience*, **86**, 1235–1244 (1998).
20. Li, B.Y., and Schild, J.H., Patch clamp electrophysiology in the nodose ganglia of the adult rat, *J. Neurosci. Meth.*, **115**, 157–167 (2002).
21. Schild, J.H., and Kunze, D.L., Experimental and modeling study of Na+ current heterogeneity in rat nodose neurons and its impact on neuronal discharge, *J. Neurophysiol.*, **78**, 3198–3209 (1997).

22. Lawson, S.N., Morphological and biochemical cell types of sensory neurons. In: *Sensory Neurons: Diversity, Development, and Plasticity*, edited by S.A. Scott (Oxford University Press, New York, 1992), pp. 27–59.
23. Amaya, F., Decosterd, I., Samad, T.A., Plumpton, C., Tate, S., Mannion, R.J., Costigan, M., and Woolf, C.J., Diversity of expression of the sensory neuron-specific TTX-resistant voltage-gated sodium ion channels SNS and SNS2, *Mol. Cell Neurosci.*, **15**, 331–342 (2000).
24. Fitzgerald, M., Capsaicin and sensory neurons: a review, *Pain*, **15**, 109–130 (1983).
25. Browning, K.N., Excitability of nodose ganglion cells and their role in vago-vagal reflex control of gastrointestinal function, *Curr. Opin. Pharmacol.*, **3**, 613–617 (2003).
26. Schelegle, E.S., Chen, A.T., and Loh, C.Y., Effects of vagal perineural capsaicin treatment on vagal efferent and airway neurogenic responses in anesthetized rats, *J. Basic Clin. Physiol Pharmacol.*, **11**, 1–16 (2000).
27. Fan, W., Reynolds, P.J., and Andresen, M.C., Baroreflex frequency-response characteristics to aortic depressor and carotid sinus nerve stimulation in rats, *Am. J. Physiol.*, **271**, H2218–H2227 (1996).
28. Sapru, H.N., Gonzalez, E., and Krieger, A.J., Aortic nerve stimulation in the rat: cardiovascular and respiratory responses, *Brain Res. Bull.*, **6**, 393–398 (1981).
29. Sapru, H.N., and Krieger, A.J., Carotid and aortic chemoreceptor function in the rat, *J. Appl. Physiol.*, **42,** 344–348 (1977).
30. Kobayashi, M., Cheng, Z.B., Tanaka, K., and Nosaka, S., Is the aortic depressor nerve involved in arterial chemoreflexes in rats? *J. Auton. Nerv. Syst.*, **78**, 38–48 (1999).
31. De Paula, P.M., Castania, J.A., Bonagamba, L.G.H., Salgado, H.C., and Machado, B.H., Hemodynamic responses to electrical stimulation of the aortic depressor nerve in awake rats, *Am. J. Physiol.*, **277,** R31–R38 (1999).
32. Fan, W., and Andresen, M.C., Differential frequency-dependent reflex integration of myelinated and nonmyelinated rat aortic baroreceptors, *Am. J. Physiol.*, **275**, H632–H640 (1998).
33. Andresen, M.C., Krauhs, J.M., and Brown, A.M., Relationship of aortic wall baroreceptor properties during development in normotensive and spontaneously hypertensive rats, *Circ. Res.*, **43**, 728–738 (1978).
34. Fazan, V.P., Salgado, H.C., and Barreira, A.A., Aortic depressor nerve unmyelinated fibers in spontaneously hypertensive rats, *AJP—Heart and Circulatory Physiology*, **280,** H1560–H1564 (2001).
35. Krauhs, J.M., Structure of rat aortic baroreceptors and their relationship to connective tissue, *J. Neurocyt.*, **8**, 401–414 (1979).
36. Doan, T.N., Stephans, K., Ramirez, A.N., Glazebrook, P.A., Andresen, M.C., and Kunze, D.L., Differential distribution and function of hyperpolarization-activated channels in sensory neurons and mechanosensitive fibers, *J. Neurosci.*, **24**, 3335–3343 (2004).
37. Schild, J.H., Clark, J.W., Hay, M., Mendelowitz, D., Andresen, M.C., and Kunze, D.L., A- and C-type nodose sensory neurons: model interpretations of dynamic discharge characteristics, *J. Neurophysiol.*, **71**, 2338–2358 (1994).
38. Andresen, M.C., and Yang, M., Arterial baroreceptor resetting: contributions of chronic and acute processes, *Clin. Exper. Pharmacol. Physiol.*, **15** (suppl.), 19–30 (1989).

39. Kunze, D.L, and Andresen, M.C., Arterial baroreceptors: excitation and modulation. In: *Reflex Control of the Circulation*, edited by I.H. Zucker and J.P. Gilmore (CRC Press, Boca Raton, FL, 1991), pp. 141–166.
40. Douglas, W.W., Ritchie, J.M., and Schaumann, W., Depressor reflexes from medullated and nonmedullated fibres in the rabbit's aortic nerve, *J. Physiol.*, **132**, 187–198 (1956).
41. Douglas, W.W., and Ritchie, J.M., Cardiovascular reflexes produced by electrical excitation of non-medullated afferents in the vagus, carotid sinus and aortic nerves, *J. Physiol*, **134,** 167–178 (1956).
42. Douglas, W.W., Ritchie, J.M., and Schaumann, W., A study of the effect of the pattern of electrical stimulation of the aortic nerve on the reflex depressor responses., *J. Physiol.*, **133**, 232–242 (1956).
43. Fan, W., Schild, J.H., and Andresen, M.C., Graded and dynamic reflex summation of myelinated and unmyelinated rat aortic baroreceptors, *Am. J. Physiol.*, **277**, R748–R756 (1999).
44. Loewy, A.D., Central autonomic pathways. In: *Central Regulation of Autonomic Functions*, edited by A.D. Loewy and K.M. Spyer (Oxford University Press, New York, 1990), pp. 88–103.
45. Spyer, K.M., The central nervous organization of reflex circulatory control. In: *Central Regulation of Autonomic Functions*, edited by A.D. Loewy and K.M. Spyer (Oxford University Press, New York, 1990), pp. 168–188.
46. Pilowsky, P.M., and Goodchild, A.K., Baroreceptor reflex pathways and neurotransmitters: 10 years on, *J. Hypertens.*, **20**, 1675–1688 (2002).
47. Andresen, M.C., Kunze, D.L., and Mendelowitz, D., Central nervous system regulation of the heart. In: *Basic and Clinical Neurocardiology*, edited by J.A. Armour and J.L. Ardell (Oxford University Press, New York, 2004), pp. 187–219.
48. Mendelowitz, D., Yang, M., Andresen, M.C., and Kunze, D.L., Localization and retention in vitro of fluorescently labeled aortic baroreceptor terminals on neurons from the nucleus tractus solitarius, *Brain Res.*, **581**, 339–343 (1992).
49. Ciriello, J., Brainstem projections of aortic baroreceptor afferent fibers in the rat, *Neurosci. Lett.*, **36**, 37–42. (1983).
50. Kalia, M., and Welles, R., Brain stem projections of the aortic nerve in the cat: a study using tetramethyl benzidine as the substrate for horseradish peroxidase, *Brain Res.*, **188**, 23–32 (1980).
51. Gao, X., Phillips, P.A., Widdop, R.E., Trinder, D., Jarrott, B., and Johnston, C.I., Presence of functional vasopressin V1 receptors in rat vagal afferent neurones, *Neurosci. Lett.*, **145,** 79–82 (1992).
52. Ding,Y.Q. , Li, J.L., Lü, B.Z., Wang, D., Zhang, M.L., and Li, J.S., Co-localization of *m*-opioid receptor-like immunoreactivity with substance P-LI, calcitonin gene-related peptide-LI and nitric oxide synthase-LI in vagal and glossopharyngeal afferent neurons of the rat, *Brain Res.*, **792**, 149–153 (1998).
53. Hoang, C.J., and Hay, M., Expression of metabotropic glutamate receptors in nodose ganglia and the nucleus of the solitary tract, *Am. J. Physiol. Heart Circ. Physiol.*, **281**, H457–H462 (2001).
54. Collingridge, G.L., The brain slice preparation: a tribute to the pioneer Henry McIlwain, *J. Neurosci. Methods*, **59**, 5–9 (1995).
55. Scheuer, D.A., Zhang, J., Toney, G.M., and Mifflin, S.W., Temporal processing of aortic nerve evoked activity in the nucleus of the solitary tract, *J. Neurophysiol.*, **76,** 3750–3757 (1996).

56. Czachurski, J., Dembowsky, K., Seller, H., Nobling, R., and Taugner, R., Morphology of electrophysiologically identified baroreceptor afferents and second-order neurons in the brainstem of the cat, *Arch. Ital. Biol.*, **126**, 129–144 (1988).
57. Donoghue, S., Felder, R.B., Gilbey, M.P., Jordan, D., and Spyer, K.M., Post-synaptic activity evoked in the nucleus tractus solitarius by carotid sinus and aortic nerve afferents in the cat, *J. Physiol.*, **360**, 261–273 (1985).
58. Mifflin, S.W., and Felder, R.B., An intracellular study of time-dependent cardiovascular afferent interactions in nucleus tractus solitarius, *J. Neurophysiol.*, **59**, 1798–1813 (1988).
59. Miura, M., Post-synaptic potentials recorded from nucleus of the solitary tract and its subjacent reticular formation elicited by stimulation of the carotid sinus nerve, *Brain Res.*, **100**, 437–440 (1975).
60. Doyle, M.W., Bailey, T.W., Jin, Y.H., Appleyard, S.M., Low, M.J., and Andresen, M.C., Strategies for cellular identification in nucleus tractus solitarius slices, *J. Neurosci. Methods*, **37**, 37–48 (2004).
61. Andresen, M.C., and Yang, M., Non-NMDA receptors mediate sensory afferent synaptic transmission in medial nucleus tractus solitarius, *Am. J. Physiol.*, **259**, H1307–H1311 (1990).
62. Miles, R., Frequency dependence of synaptic transmission in nucleus of the solitary tract in vitro, *J. Neurophysiol.*, **55**, 1076–1090 (1986).
63. Doyle, M.W., and Andresen, M.C., Reliability of monosynaptic transmission in brain stem neurons in vitro, *J. Neurophysiol.*, **85**, 2213–2223 (2001).
64. Doyle, M.W., Bailey, T.W., Jin, Y.-H, and Andresen, M.C., Vanilloid receptors presynaptically modulate visceral afferent synaptic transmission in nucleus tractus solitarius, *J. Neurosci.*, **22**, 8222–8229 (2002).
65. Drewe, J.A., Childs, G.V., and Kunze, D.L., Synaptic transmission between dissociated adult mammalian neurons and attached synaptic boutons, *Science*, **241**, 1810–1813 (1988).
66. Jin, Y.-H., Bailey, T.W., Li, B.Y., Schild, J.H., and Andresen, M.C., Purinergic and vanilloid receptor activation releases glutamate from separate cranial afferent terminals., *J. Neurosci.*, **24**, 4709–4717 (2004).
67. Moran, M.M., Xu, H., and Clapham, D.E., TRP ion channels in the nervous system, *Curr. Opin. Neurobiol.*, **14**, 362–369 (2004).
68. Mendelowitz, D., and Kunze, D.L., Identification and dissociation of cardiovascular neurons from the medulla for patch clamp analysis, *Neurosci. Lett.*, **132**, 217–221 (1991).
69. Jin, Y.-H., Bailey, T.W., Doyle, M.W., Li, B.Y., Chang, K.S.K., Schild, J.H., Mendelowitz, D., and Andresen, M.C., Ketamine differentially blocks sensory afferent synaptic transmission in medial nucleus tractus solitarius (mNTS), *Anesthesiology*, **98**, 121–132 (2003).
70. Borodinsky, L.N., Root, C.M., Cronin, J.A., Sann, S.B., Gu, X., and Spitzer, N.C., Activity-dependent homeostatic specification of transmitter expression in embryonic neurons, *Nature*, **429**, 523–530 (2004).
71. Andresen, M.C., Doyle, M.W., Bailey, T.W., and Jin, Y.-H., Differentiation of autonomic reflex control begins with cellular mechanisms at the first synapse within the nucleus tractus solitarius, *Braz. J. Med. Biol. Res.*, **37**, 549–558 (2004).
72. Jin, Y.-H., Bailey, T.W., and Andresen, M.C., Cranial afferent glutamate heterosynaptically modulates GABA release onto second order neurons via distinctly segregated mGluRs. *J. Neurosci.*, (2004).

73. Benham, C.D., Davis, J.B., and Randall, A.D., Vanilloid and TRP channels: a family of lipid-gated cation channels, *Neuropharmacology*, **42**, 873–888 (2002).
74. Caterina, M.J., Vanilloid receptors take a TRP beyond the sensory afferent, *Pain*, **105**, 5–9 (2003).
75. Kumada, M., Terui, N., and Kuwaki, T., Arterial baroreceptor reflex: its central and peripheral neural mechanisms, *Prog. Neurobiol.*, **35**, 331–361 (1990).
76. Mendelowitz, D, Firing properties of identified parasympathetic cardiac neurons in nucleus ambiguus. *Am. J. Physiol.*, **271**(6), H2609–H2614 (1996).
77. Neff, R.A., Mihalevich, M., and Mendelowitz, D., Stimulation of NTS activates NMDA and non-NMDA receptors in rat cardiac vagal neurons in the nucleus ambiguus. *Brain Res.*, **792**(2), 277–282 (1998).
78. Willis, A., Mihalevich, M., Neff, R.A., and Mendelowitz, D., Three types of postsynaptic glutamatergic receptors are activated in DMNX neurons upon stimulation of NTS. *Am. J. Physiol.*, **271**(6), R1614–R1619 (1996).

6
TRPV1 as a Molecular Transducer for Salt and Water Homeostasis

Donna H. Wang* and Jeffrey R. Sachs

6.1. Introduction and Overview of the TRP Family and TRPV1

The transient receptor potential (TRP) family of ion channels was first characterized in *Drosophila*, where the *trp* gene was found to be required for visual transduction in a phospholipase C (PLC) dependent process.[1] This was the first of what is now a superfamily of TRP channels that is composed of 39 channel subunit genes divided into 7 subfamilies.[2] These subfamilies include TRPC (canonical), TRPV (vanilloid), TRPM (melastatin), TRPML (mucolipin), TRPA (ankyrin), TRPP (polycystin), and TRPN (NOMP). While each subfamily has at least one vertebrate representative, all but TRPN can claim a mammalian representative.[1,3] In general, the TRP superfamily of ion channels represents a diverse set of proteins whose main function is the regulation of the plasma membrane's ion permeability.[4]

All TRP channels bear some level of sequence homology, cation selectivity, and basic channel architecture. This basic subunit consists of six transmembrane (TM) domains with a pore domain between the fifth (S5) and sixth (S6) domains. A seventh N-terminal hydrophobic domain (h1) has been predicted by hydropathy analysis to be present in many TRPC, TRPV, TRPM, TRPML, and TRPP family members.[2] Like voltage-gated K^+ channels, it is likely that TRP channels homo- or heterotetramerize into channel complexes. The subunit composition of the resulting TRP channel may directly influence its biophysical and regulatory properties.[5]

Together, the TRP channels regulate a wide variety of physiological functions. Their main importance is in sensory physiology, playing roles in smell, taste, vision, hearing, mechanosensation, and thermosensation. TRP channel functions are most certainly not limited to sensation, however. Indeed there is remarkable diversity in function even within a given subfamily. TRPC family channels have been implicated in functions as wide ranging as the acrosome reaction (TRPC2)

*Department of Medicine, Neuroscience, and Cell and Molecular Biology Program, B316B Clinical Center, Michigan State University, East Lansing, MI 48824-1313; e-mail: Donna.Wang@ht.msu.edu to modulation of neurite extension (TRPC5). Those in the TRPV subfamily are mainly implicated in thermosensation (TRPV1-4),[6] but have also been implicated in osmosensation (TRPV4)[7] and calcium reabsorption in the kidney and the GI tract (TRPV5, 6).[8] Channels in the TRPM subfamily have been identified with Mg^{2+} absorption (TRPM6) and with sensing cool temperatures (TRPM8).[9] TRPA channels have been associated with cold pain sensation (TRPA1), while TRPN channels function in hearing (TRPN1). Polycystic kidney disease has been localized to a defect in a TRPP channel (TRPP2), whereas mucolipidosis is associated with dysfunction of a TRPML channel (TRPML). Comprehensive reviews of these and other functions of all of the TRP channels are reported elsewhere.[1–3,10–13]

TRPV1, a recently cloned member of the TRPV subfamily, has well-recognized roles in pain[14] and thermosensation.[15] Findings also indicate that TRPV1 participates in particulate matter-induced apoptosis[16], normal bladder function,[17] taste,[18] and neurogenic inflammation.[19] Intense study has drawn attention to TRPV1's potential role in the regulation of salt and water homeostasis and its subsequent effect on systemic blood pressure. In addition to providing an overview of TRPV1 function, activation, and regulation, the focus of this chapter will be the role of TRPV1 in the cardiovascular system, particularly in the regulation of salt sensitivity of arterial pressure.

Members of the TRPV family are known to be activated by a wide array of stimuli, including acidity, heat, lipids, phosphorylation, phorbol esters, changes in osmolarity, and others. The TRPV channels are the only members of the TRP superfamily known to be activated by vanilloids, the property for which the subfamily was named. Because the vanilloid capsaicin is a known activator of TRPV1, the protein is also known as the "capsaicin receptor." TRPV1 can be activated by multiple stimuli,[20,21] and as a result it has been hypothesized that it may function as a "molecular integrator" of biological systems. This hypothesis stems from the observation that protons can lower the threshold for heat activation of TRPV1.[22]

As a result of its complex modes of activation, much study has been done on the molecular structure of TRPV1 and its biochemical characterization.[23] When the gene encoding TRPV1 was initially cloned in 1997, it was discovered that TRPV1 cDNA contains a 2,514 nucleotide open reading frame encoding an 838 amino acid peptide with a molecular mass of around 95 kDa.[14] Both termini point intracellularly. TRPV1 was described as having an N-terminus of 432 amino acids that notably contains three ankyrin repeat domains following a proline-rich region. These ankyrin repeats are believed to function as an interconnection of the TRPV1 protein with spectrin-based cytoskeletal elements.[24] Furthermore, it has been discovered that calmodulin (CaM) binds to the first ankyrin repeat (residues 189-222).[25] The C-terminus was found to consist of 154 amino acids, although with no currently recognizable motifs. The C-terminus has been determined, however, to contain domains responsible for allosteric conformational changes that occur after ligand binding.[26] Further study has indicated that the C-terminus also contains a 35-amino acid segment (residues 767-801) that is bound by CaM.[27]

It is likely that the native quaternary structure of the TRPV1 receptor is composed of four equivalent 95 kDa subunits.[28] This evidence was supported by

Kuzhikandathil et al. when they developed a TRPV1 subunit with a dominant negative mutation through mutation of residues in the sixth transmembrane domain.[29] This resulted in a receptor that was unable to be activated by capsaicin. Molecular determinants of the association domain of TRPV1 were described in 2004 by Garcia-Sanz et al.[30] In this study they found that a stretch of amino acids on the C-terminus spanning Glu^{684} to Arg^{721} composes a TRP-like domain similar to the 25-amino acid TRP domains highly conserved in many other TRP channels.[12] Deletion of the TRP-like domain from the C-terminus of one TRPV1 subunit significantly inhibited TRPV1 receptor activation by vanilloids, indicating that the TRP-like domain is an association domain that is necessary for the formation of a fully functional receptor. Given that a deletion mutation can negatively regulate TRPV1 receptor formation, it is not surprising that genomic studies have provided evidence for several splice variant TRPV1 cDNAs that could modulate TRPV1 function.[31]

When a homomultimer of functional TRPV1 subunits forms, the result is a TRPV1 receptor that functions as a nonselective cation channel that exhibits a time- and Ca^{2+}-dependent outward rectification followed by a long-lasting refractory period.[32] In terms of cation selectivity, TRPV1 has been shown to have no preference for monovalent cations, but out of divalent cations it prefers calcium over magnesium.[14] Thus, when TRPV1 is activated the pore of the channel opens resulting in an influx of extracellular calcium that effects a cellular response.

6.2. Expression of TRPV1

The functions of TRPV1 are dependent on its localization within the body and expression within the cell. In the original report by Caterina et al., TRPV1 was reported to be exclusively localized to primary sensory neurons in the dorsal root ganglia (DRG).[14] In this study evidence indicated that TRPV1 is expressed mainly in small neurons with unmyelinated C fibers. In 2002, a study indicated that TRPV1 is also expressed in large neurons with myelinated Aδ fibers.[33] A more recent study, however, has suggested that TRPV1 is found mainly in the unmyelinated C fibers.[34] Furthermore, these DRG neurons have been classified based on coexpression of TRPV1 with receptors for neurotrophic factors.[35] The majority of neurons that express trkA, IB4,[36] SP, and CGRP are TRPV1-positive neurons, although the exact phenotype of these neurons varies across species.[37] These primary afferent neurons innervate a wide variety of tissues, including virtually all vascular beds. In addition to the dorsal root ganglia, TRPV1 has been found in the trigeminal and nodose ganglia.

Since 1997, TRPV1 has been found to be present in many non-neuronal tissues in both rats and humans. The ever-growing list of organs of expression of TRPV1 has led to the discovery of novel functions of the TRPV1 receptor that has implications for health and disease.[38] In addition to its expression in DRG neurons and the CNS,[39] studies with rat have detected TRPV1 protein and/or mRNA in the kidney,[39] the bladder,[40] urothelium,[40,41] heart,[42,43] stomach,[44] mast cells,[45] pulmonary arterial and aortic smooth muscle,[46] and spleen.[39] Careful attention must be given in interpreting the results of such studies as methods such as reverse transcriptase-polymerase chain reaction (RT-PCR) can be rather imprecise in locating TRPV1 mRNA. For instance, in the kidney it is difficult to distinguish between TRPV1 mRNA located in the epithelial cells, vascular smooth muscle or endothelial cells, or the sensory nerves innervating these effector structures, especially if a chunk of tissues is used for analysis.

The rapidly growing list of human tissues that express TRPV1 currently includes human epidermal keratinocytes,[47,48] mast cells,[49] epithelial cells of hair follicle,[48] sweat glands,[48] sebaceous glands,[48] bladder urothelium,[50] kidney,[51] cerebral cortex,[51] cerebellum,[51] and hypothalamus,[52] among others. Thus, TRPV1 has been implicated widely in physiology and pathology, as comprehensively reviewed elsewhere.[53,54]

6.3. Agonists and Antagonists of TRPV1

TRPV1's wide expression suggests that it has a complex mode of activation and regulation, and indeed this is the case. Consistent with the suggestion that TRPV1 is a polymodal integrator of noxious stimuli, numerous biological and synthetic compounds have been discovered that can activate the receptor. Three categories of TRPV1 activation include: receptor activation, ligand activation, and direct activation.[2] Receptor activation refers to the activation of isoforms of phospholipase C through the activity of G-protein-coupled receptors and receptor tyrosine-kinases. As a result, phosphotidylinositol-4,5-bisphosphate (PIP_2) is hydrolyzed into diacylglycerol and inositol-3,4,5-trisphosphate (IP_3), products that can modulate TRPV1 function. Furthermore, specific isoforms of PKA and PKC are activated that can modulate the channel's status through phosphorylation events. Ligand activation, the most commonly studied form of activation of TRPV1 proteins, refers to the binding of either exogenous or endogenous small organic molecules, inorganic ions (such as H^+), or products of lipid or nucleotide metabolism to the channel (either extra- or intracellularly) in a way that causes a conformational change in the channel that opens the pore to allow influx of cations. Finally, direct activation refers to mechanical stimuli or changes in temperature.

TRPV1 was the founding member of the vanilloid subfamily of TRP subunits because it can be activated by molecules with a vanillyl moiety, including capsaicin, olvanil, and others. Capsaicin and anandamide (arachidonoyl ethanolamide), a vanilloid-like compound that is a potent agonist at the TRPV1 receptor,[55] have been most widely studied and used in the biochemical and pharmacological characterization of the receptor.

Anandamide was discovered as an agonist of the TRPV1 receptor when patch-clamp experiments expressing TRPV1 exhibited anandamide-induced currents in whole cells and isolated membrane patches.[56] Anandamide was later described as a full agonist at the human TRPV1 receptor when electrophysiological experiments showed that both capsaicin and anandamide induced similar Ca^{2+}-mediated inward currents in hTRPV1 transfected HEK293 cells.[57] Later it was shown that anandamide can only produce a Ca^{2+}-mediated current in primary cultures of DRG neurons at a pH ≤ 6.5.[58] Much interest has surrounded anandamide as a potential candidate for the endogenous activator of TRPV1 proteins, but the quest to discover a mechanism for how endogenous anandamide activates TRPV1 has proved difficult. A recent study provides evidence that anandamide is formed in cells following activation of the PLC/IP_3 pathway by a rise in intracellular calcium.[59] Thus, anandamide may function as a second messenger inside the cell that amplifies calcium levels via TRPV1 by sensing the calcium release from intracellular stores. This hypothesis remains to be confirmed.

Many other ligands have been discovered and synthesized that are activators of TRPV1. These include but are not limited to methanandamide,[60,61] N-arachidonoyl-dopamine (NADA),[62,63] resiniferatoxin (RTX), rinvanil and its derivatives,[64] 2-aminoethoxydiphenyl borate (2-APB),[65,66] the peripheral satiety factor oleoylethanolamide,[67] N-oleoyldopamine,[68] ethanol,[69–71] as well as several lipoxygenase products[72]: 12-(S)-HPETE, 15-(S)-HPETE, 5-(S)-HETE, 15-(S)-HETE, leukotriene B4, and a cytochrome P450 product 20-HETE.[73]

Beyond these ligand-mediated activations, TRPV1 has been shown to be subject to direct activation by heat (elevated temperatures $\geq 43°C$).[14,74] Furthermore, protons have been shown to lower the threshold for direct activation of TRPV1 by heat[22] and for receptor activation mediated by capsaicin.[75]

Similarly, ATP has been shown to potentiate TRPV1 activity through its interaction with P2Y receptors in a PKC-dependent pathway.[76] As a result of this potentiation the threshold for TRPV1 activation is lowered to room temperature. Another example of receptor activation of TRPV1 was discovered recently. It was found that 1,2-napthoquinone causes the phosphorylation of protein tyrosine kinases, leading to the activation of the phospholipase A2/lipoxygenase signaling pathway that causes the TRPV1-dependent contraction of guinea pig tracheal smooth muscle.[77]

Similarly, several compounds have been identified that antagonize TRPV1 receptor activation. These include the widely studied capsazepine and ruthenium red (RR),[78] as well as the high-affinity iodo-resiniferatoxin,[79] thiazole carboxamides,[80] A-425619,[81,82] AMG 9810,[83] and N-acylvanillamines such as arvanil.[84] Together these represent a class of compounds that is known to inhibit TRPV1 channel activation.

TRPV1 antagonists have come under intense investigation due to their therapeutic promise in the treatment of pain.[85] Thus, much effort has gone into understanding the molecular determinants of receptor agonism and antagonism as exhibited by the various modulators of TRPV1 function.

The discovery that vanilloids bind to TRPV1 from the intracellular side led investigators to develop deletion and site-directed mutagenesis experiments on the cytosolic tails of both termini to determine ligand recognition sites. Such experiments determined that Glu-761 and Arg-114 are recognition sites for capsaicin binding in rat TRPV1.[86] When the two mutants were co-expressed, capsaicin elicited no inward current, indicating that vanilloids interact with both termini of the TRPV1 subunit. Gating by capsaicin is notably more complex, however, as indicated by the potentiation of channel opening by heat and acid. A molecular determinant for the potentiation of capsaicin activation by acid was localized to Glu-600.[87] These same authors demonstrated that acid was enough to activate the receptor itself in a way distinct from other forms of activation, as mutations at Glu-648 selectively destroyed proton-evoked activation without affecting channel responses to vanilloids or heat.

A study that utilized rabbit TRPV1 was able to determine key residues in transmembrane regions 3 and 4 that are critical for vanilloid binding. Mutations in Met-547 and Thr-550 were able to confer vanilloid sensitivity (both capsaicin and RTX) to rabbit TRPV1 that is normally 100-fold less sensitive to vanilloids than either rat or human.[88] These findings add to the previously reported finding that Tyr-511 and Ser-512 are critical residues for conferring vanilloid sensitivity in TRPV1.[89] Another study found that mutating the Ser-512 equivalent residue in human TRPV1 destroyed vanilloid-dependent activation of the channel even though capsaicin was still able to interact with the mutant residue.[90] Thus, it is evident that several distinct areas of the channel are involved in its gating by vanilloid ligands. The interested reader is directed to a more comprehensive review on the molecular determinants of vanilloid binding.[23]

In addition to TRPV1's complex interaction with vanilloid ligands, it has been reported that distinct mechanisms exist for TRPV1 activation by either heat or acid.[75] Recently reported indirect evidence indicates that heat causes a conformational change in the receptor. This can be concluded from the finding that the potent TRPV1 antagonist iodoresiniferatoxin (IRTX) prevented heat-evoked responses in rats.[91] Understanding how agonists and antagonists interact with the receptor is an important area of research for those interested in drug design.

6.4. TRPV1 Signaling Pathways

Beyond regulation at the gate of the channel, TRPV1 is known to be regulated by a variety of intracellular signaling pathways. The actions of several protein kinases and phosphatases work in concert to determine the activation status of the channel. These and other mechanisms of TRPV1 regulation will now be briefly reviewed.

In 2000, it was discovered that TRPV1 channel activity could be induced by the activation of protein kinase C (PKC), and that both bradykinin and anandamide enhanced channel activity in a PKC-dependent manner.[92] PKC has also been implicated in re-sensitization of the receptor after desensitization was induced by repeated capsaicin treatment.[93] Later studies used the phorbol ester PMA (an activator of intracellular PKC) to identify Ser-502 and Ser-800 as key target residues of PKCε phosphorylation on rat TRPV1's first intracellular loop and C-terminus, respectively.[94] These results were expanded upon in a Bhave et al. study in which *in vitro* phosphorylation and protein sequencing techniques were used to add Thr-704 from the C-terminus and Thr-144 from the N-terminus to the list of major PKC phosphorylation sites.[95] PKC's role in receptor desensitization is not limited to phosphorylation status, however. Deletion studies of rat TRPV1 have indicated that the distal C-terminus plays an inhibitory role in counteracting PKC stimulation.[96]

Knowledge of the role of PKC on TRPV1 activation led to the search for ways in which PKC could be activated in the physiologic state, particularly during inflammation. Chuang et al. suggested that both of the inflammatory mediators bradykinin and nerve growth factor (NGF) modulate TRPV1 sensitivity through both PKC-dependent and independent pathways.[97] Both bradykinin and NGF stimulate phospholipase C to cleave PIP_2 into IP_3 and DAG. Consequences of this event are twofold. First, IP_3 causes the release of calcium from intracellular stores. Increased intracellular calcium then works in concert with DAG to activate PKC, which can subsequently stimulate TRPV1. Furthermore, a more recent study indicated that PLC-dependent potentiation of TRPV1 depends on residues 777-820, a PIP_2binding site on the C-terminus of the receptor.[98] Thus, activation of PLC can relieve TRPV1 from PIP_2-mediate inhibition.

In addition to its role in directly mediating receptor activity, PKC has been implicated in the trafficking of vesicular channels containing TRPV1 to the cell surface via SNARE-dependent exocytosis.[99] These authors suggested that this works through the rapid recruitment of the SNARE proteins Snapin and Syt IX, two proteins identified to interact with the ankyrin repeats on the N-terminus of TRPV1. This represents a form of indirect regulation that can sensitize or desensitize the channel to stimulation by a specific agonist molecule.[2]

These and other studies indicated that PKC signaling may be a mechanism for transducing environmental stimuli into TRPV1 activation. The activation of TRPV1 via PKC-dependent pathways provides a mechanism through which inflammatory mediators such as nerve growth factor, bradykinin, and the proinflammatory cytokine IL-1β can contribute to TRPV1-mediated inflammatory hyperalgesia.[100–102] Future studies can be expected to implicate other endogenous inflammatory mediators in this TRPV1-mediated pathway.

PKC is not the only intracellular kinase involved in modulation of TRPV1 activity. A 1998 study found that prostaglandin E_2 enhances the gating of these channels by capsaicin via the cAMP-PKA signaling pathway.[103] A more recent study found that PKA inhibits the capsaicin-evoked desensitization of the channel by phosphorylating Ser-116 and Thr-370 residues.[104] In addition, Ser-502 has been discovered as a PKA phosphorylation site.[105] The importance of PKA in this event is affirmed by studies that inhibited calcineurin, a phosphatase found to regulate the desensitization of the capsaicin-activated channel.[106] When the cyclosporine A·cyclophilin A complex, a specific inhibitor of calcineurin, was applied intracellularly, a decrease in capsaicin-mediated desensitization was observed similar to when intracellular PKA was activated.[107]

Also regulating the capsaicin-mediated calcium-dependent desensitization is the Ca^{2+}-calmodulin dependent kinase II (CaMKII). In 2004, a study found that mutations at two consensus sites for this kinase failed to confer a capsaicin-sensitive current.[108] Phosphorylation by PKA and PKC converge at these same sites. Together, these results indicate that PKA, PKC, and CaMKII work in opposition to calcineurin to regulate the calcium-dependent desensitization of TRPV1.

This aforementioned regulation of TRPV1 by phosphorylation and dephosphorylation are not the only physiologically relevant methods of regulation. Several TRPV1 splice variants have been isolated in mouse and rat models that can modulate TRPV1-mediated responses to environmental stimuli. The mouse variant TRPV1β encodes a dominant-negative subunit of the TRPV1 channel such that it inhibits channel activity when it associates with other normal TRPV1 subunits in a tetramer.[109] Furthermore, a novel splice variant was recently discovered in the rat renal papilla that could add another point of regulation to the TRPV1-mediated response to a salt load.[110]

As mentioned earlier, the C-terminus of TRPV1 is known to contain a 35-amino acid segment that is bound by calmodulin.[27] These authors found that when this segment is interrupted, capsaicin-mediated desensitization is prevented. Calmodulin may be part of a feedback mechanism through which calcium entering the cell as a result of receptor activation binds to CaM, inhibiting channel gating.[111] Both of the intracellular binding domains for CaM exhibit some binding in the absence of calcium, from which it can be inferred that some CaM is present on TRPV1 at all times. Calmodulin's role in the complex regulation of TRPV1 remains to be fully elucidated.

Other noteworthy regulation of TRPV1 function includes the increased expression of TRPV1 receptors as mediated by the inflammatory mediators ATP, bradykinin, and NGF,[112] regulation of mean open and closed channel times by the association of Fas-associated factor 1 (FAF1) with the N-terminus of TRPV1,[113] the mobilization of TRPV1 to the plasma membrane in the presence of insulin,[114] the regulation of the oxidation state of key cysteine residues on the extracellular side of the receptor by dithiothreitol,[115] and the upregulation of mRNA expression by neurotrophic factors.[35] As a result of this complex web of regulatory mechanisms there is tremendous potential for future study into how these systems interact in many physiological and pathophysiological states.

6.5. The Function of TRPV1 as a Chemo- and Mechanosensor

The polymodal activation of TRPV1 by such noxious stimuli as high temperatures and acidity have indicated that one of its major functions is to act as a molecular transducer of a painful physico-chemical environment. Beyond nociception, the role of TRPV1 has expanded to other aspects of physiological regulation, notably the cardiovascular system. A prominent example involves the role of TRPV1 in the antihypertensive mechanisms induced by sodium loading.[116] One prominent hypothesis is that the release of the vasodilatory neuropeptides substance P (SP) and calcitonin gene-related peptide (CGRP) from terminals of sensory neurons innervating the vascular beds compensates for a rise in systemic blood pressure.[117] It is currently unknown, however, how increases in plasma/interstitial sodium concentrations or intralumenal pressure lead to activation of TRPV1. Thus, it is of great interest to understand how the receptor senses the chemical and mechanical environment in the vasculature and kidney and integrates altered environmental stimuli to lead to compensatory responses in a pathophysiological state such as hypertension.

It is apparent from the previous discussion that TRPV1 has the potential to act as a sensor of a wide range of chemical stimuli. The question remains, however, as to what are the endogenous ligands of TRPV1 in the CNS and peripheral sites. At least three to four categories of candidates have been identified: anandamide, lipoxygenase and cytochrome P450 products of arachidonic acid, and N-arachidonoyldopamine.[73,118] While much effort has been put forth to understand the biosynthesis and degradation of each of these possible endogenous ligands, little data exists on the role of each in activating TRPV1 *in vivo*. Thus, future studies will be focused on determining the roles of each in the pathophysiological state, a process that in all likelihood will lead to the discovery of new candidates for endogenous activation of TRPV1.

Evidence regarding the role of TRPV1 as a mechanosensor is, at best, inconclusive. No evidence is available supporting the direct activation of TRPV1 via sheer stress generated by the flow in the lumen of blood vessels. However, TRPV1 may be activated indirectly by altered transmural pressure through the production of 20-hydroxyeicosatetraenoic acid (20-HETE), a known activator of TRPV1 that causes the release of SP when it binds to the TRPV1 expressed on sensory C-fibers.[73] The same authors concluded that TRPV1 was not a mechanosensor, however, because the stretch-activated nonselective cation channel blocker gadolinium had no effect on capsaicin-induced activation of TRPV1.

On the contrary, indirect evidence indicates that perhaps TRPV1 is a mechanosensor. As a result of the subsequent increase in cellular volume, an osmotic stimulus leads to changes in membrane tension that can be regarded as a mechanical stimulus. When TRPV1 null mice were studied, it was found that the bladder epithelial cells did not respond to a hypotonic osmotic stimulus.[17] Wild type bladder urothelial cells, however, responded to a hypotonic stimulus with a release of ATP. Thus, the authors concluded that the absence of TRPV1 in the urothelium inhibits normal bladder function and the normal mechanically evoked purinergic signaling by the urothelium. It may also be possible that TRPV1 co-assembles with distinct mechanosensitive TRPV subtypes or other proteins in the bladder.[34]

A study by Naeini et al. implicates TRPV1 in the osmosensory transduction mediated by arginine-vasopressin (AVP)-releasing neurons in the supraoptic nucleus (SON) of the hypothalamus.[119] In this study it was shown that N-terminal splice variants of TRPV1 are expressed by AVP neurons in the SON. In a TRPV1-positive animal, these cells would shrink in response to a hyperosmotic stimulus, activating a stretch-inhibited osmosensory transduction channel.[120] This would lead to depolarization that contributes to cellular excitation and AVP release. TRPV1 knockout mice were unable to respond to hyperosmotic stimulation, indicating that the TRPV1 gene may encode a central component of the osmoreceptor that regulates systemic levels of AVP.

6.6. The Role of TRPV1 in the Regulation of Salt and Water Homeostasis in the Physiologic State

The aforementioned study by Naeini et al. indicates that a TRPV1 splice variant in the central nervous system plays an important role in antidiuresis in response to serum hyperosmolality,[119] which appears to be opposite to the peripheral response in which TRPV1 seems to be activated by the hypotonic stimulus.[17] Thus it seems apparent that TRPV1 is an important player in the regulation of salt and water homeostasis in the physiologic state. The evidence supporting such a role for TRPV1 is shown below.

In addition to their sensory afferent function, TRPV1-positive neurons also have an efferent motor function. Binding of agonists to TRPV1 causes the opening of the cation channel that leads to the influx of sodium and calcium ions.[121–125] While sodium influx is high enough to depolarize the neuron and lead to an afferent impulse conduction, calcium influx is required for the release of neuropeptides from sensory nerve endings.[126] At least 12 different types of transmitters have been identified in TRPV1-positive sensory neurons that can be released alone or with other peptides. These neuropeptides include SP, neurokinin A, neuropeptide K, eledoisin-like peptide, somatostatin, vasoactive intestinal polypeptide, cholecystokinin-octapeptide, CGRP, galanin, corticotrophin-releasing factor, arginine vasopressin, and bombesin-like peptides.[126]

The vasodilatory neuropeptides SP and CGRP have been shown to be extremely relevant to the discussion of whole-body salt and water homeostasis. For example, they have been shown to have direct and indirect effects on tubular ion transport in the kidney, resulting in natriuretic and diuretic actions.[127–129] Furthermore, CGRP and SP are often co-localized within the nerve endings of sensory nerves found around blood vessels in virtually all vascular beds.[127,130–133] This raises the possibility that TRPV1 receptors located in sensory nerves innervating the vascular beds sense local and/or systemic chemical (e.g., changes in concentrations of ions or endocrine/paracrine/autocrine factors) or mechanical (e.g., changes in transmural pressure) stimuli and release neuropeptides such as SP and CGRP to mediate a response.

In addition, the TRPV1 agonist capsaicin has been shown to be a selective toxin for sensory neurons. It binds to TRPV1 receptors on DRG neurons causing down-regulation of the receptor, depletion of sensory neurotransmitters, and sensory nerve degeneration when given to neonates or in high doses to adults.[134] Our lab found that degeneration of sensory nerves causes a subsequent rise in systemic blood pressure in rats challenged with a salt load.[116,135–138] This implies that TRPV1 normally functions to increase natriuresis and diuresis in response to a salt challenge.

Furthermore, results from our lab in a recent study indicate that TRPV1 present in a subpopulation of sensory nerves innervating the kidney is responsible for capsaicin-induced bilateral increases in diuresis and natriuresis.[139] In this experiment capsaicin was administered either into the unilateral renal pelvis or intravenously in the presence or absence of a selective TRPV1 antagonist or unilateral renal denervation. Increases in U_{flow} and U_{Na} were observed in a dose-dependent fashion that could be abolished through blockade of TRPV1 or by ipsilateral renal denervation. Thus activation of TRPV1 in the unilateral renal pelvis resulted in an enhanced bilateral renal function via the renorenal reflex.

6.7. The Role of TRPV1 in the Regulation of Salt Sensitivity of Arterial Pressure

The previously reviewed studies indicate that TRPV1 is widely expressed in the body in areas that have a direct effect on salt and water homeostasis, notably the osmosensory neurons in the CNS, sensory nerves lining the blood vessels, and sensory nerves innervating the kidney. In combination it is likely that such expression in the body indicates a central role for TRPV1 in the regulation of blood pressure homeostasis. Understanding the interactions between blood volume and sodium levels, TRPV1, the renin-angiotensin-aldosterone system (RAAS), the sympathetic nervous system, the endothelin system, and increased oxidative stress have obvious implications with regards to the treatment of hypertension. Several studies have investigated the importance of TRPV1 with regard to hypertension, particularly in models of salt sensitivity.

Our lab demonstrated for the first time in 1998 that neonatal degeneration of TRPV1-positive sensory nerves rendered an adult rat salt-sensitive.[136] As a result, rats loaded with salt showed an impaired natriuretic and diuretic response along with an increase in blood pressure. To determine an interaction between the sensory nervous system and the RAAS or sympathetic nervous system, either losartan (a type 1 angiotensin II receptor blocker), prazosin (a selective α_1-adeno-receptor blocker), or hydralazine (a nonspecific vasodilator) was administered to high-salt plus capsaicin pretreated rats.[116] Only losartan or hydralazine, but not prazosin, prevented the development of salt-induced hypertension in this model. This implies that in sensory-intact rats there is an interaction between the sensory nervous system and the RAAS in preventing the development of salt-induced hypertension. However, because losartan and hydralazine were unable to prevent the impaired natriuretic response, it suggests that intact sensory innervation is essential for a normal natriuretic response regardless of blood pressure. It is likely that losartan and hydralazine work by vasodilatory mechanisms and not by protecting against the impaired sodium excretion in the model.

In a study designed to understand the molecular mechanisms behind the salt activation of sensory nerves, our lab tested the effects of capsaicin and capsazepine on rats fed either a high or normal sodium diet.[140] Several important findings can be taken from this study. First it was found that activation of TRPV1 by capsaicin led to an increase in plasma CGRP levels and a decrease in MAP because these actions were blocked by capsazepine. Secondly, when capsazepine was administered to rats fed a high salt diet, the pro-hypertensive effects of salt were unmasked. Furthermore, western blot analysis indicated that TRPV1 receptor expression increased in both mesenteric resistance arteries and the renal medulla in response to a high salt diet. Finally, it was shown using immunohistochemical techniques that at least a subset of TRPV1-positive sensory neurons contains CGRP because TRPV1 and CGRP co-localized in the perivascular sensory nerves innvervating the mesenteric resistance arteries. We hypothesized that the upregulation in the renal medulla could mediate an increase in the release of CGRP that would bind in a paracrine fashion to its receptors in the renal medulla where it could affect salt and water homeostasis. This key study strongly implicates TRPV1 as a normal physiological regulator of blood pressure in the face of a salt challenge.

Other studies have supported this role of TRPV1. A recent study from our lab found that the depressor effect mediated by anandamide can only be prevented when both TRPV1 and CB1 receptors are blocked.[141] This means that TRPV1 at least partially contributes to the decrease in peripheral vascular resistance caused by anandamide-induced release of CGRP, a mechanism that also operates in spontaneously hypertensive rats.[142] Furthermore, our lab found from the study of Dahl salt-sensitive rats that a lack of TRPV1 receptor activity in these genetically predisposed rats eliminated the normal counterregulatory action of sensory nerves, leaving these rats without protection in the face of a salt challenge.[143] These results have led us to develop the hypothesis that TRPV1 expressed in a genetically distinct subpopulation of primary sensory nerves is activated in response to high salt intake, which promotes natriuresis and prevents salt-induced increases in blood pressure via counterbalancing the prohypertensive systems. This hypothesis is shown schematically in Fig. 6.1.

While it is certainly an attractive hypothesis that TRPV1-induced release of CGRP normally maintains blood pressure through its local effects on peripheral resistance and its effects on tubular ion transport in the kidney, the picture is most certainly more complicated. An interaction between the sensory nervous system and the RAAS has already been discussed.[116,144,145] Furthermore, it is known that the sympathetic nervous system,[135] the endothelin system,[146,147] and oxidative stress generation[148] becomes more active upon sensory nerve degeneration plus high salt intake. The role of TRPV1 and how it interacts with these systems to prevent hypertension is comprehensively reviewed elsewhere.[149]

6.8. TRPV1 as a Target for Future Drug Development

The wide expression of TRPV1 in the body indicates that agonism and antagonism of the receptor has enormous therapeutic potential. Possible therapeutic interventions may include the use of selective blockers, down-regulation strategies such as antisense treatment, or desensitization of TRPV1.[150] One of the most

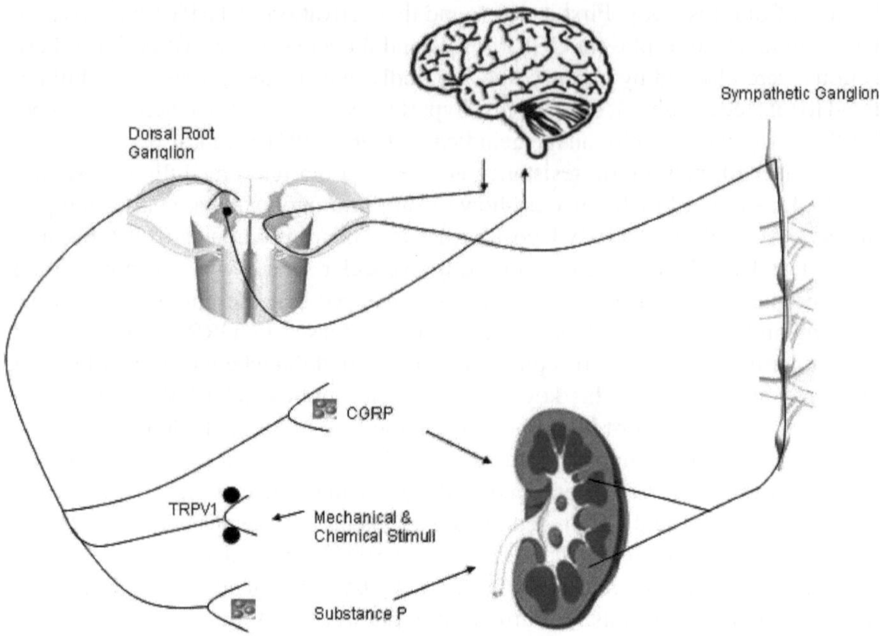

FIGURE 6.1. Activation of TRPV1 by mechanical and chemical stimuli results in the release of CGRP and SP, which promote natriuresis and diuresis through their actions on the kidney. TRPV1 also affects kidney function via descending pathways from the CNS.

widely studied of these possibilities is the use of TRPV1 antagonists in the treatment of pain.[85] Furthermore, the fact that agonists of TRPV1 are able to induce a CGRP-mediated hypotensive effect in salt hypertensive rats indicates that activating TRPV1 may be an effective means of preventing the development of hypertension[151] and its associated end organ damage. Indeed, we have shown that TRPV1 gene knockout impairs postischemic recovery in isolated perfused heart in mice,[152] indicating that TRPV1 plays a protective role against ischemic injury of the heart. This in itself is an exciting possibility considering that an estimated 65 million Americans and 1 billion subjects worldwide suffer from hypertension and its associated end organ damage. A more authoritative review on the role of TRPV1 in the development of disease has been recently published.[53]

While much progress has been made in defining the structure, activation, and functional role of TRPV1 in the physiologic state, much is still to be discovered. The quest to find safe, effective therapeutics to act at TRPV1 is currently hindered by the lack of understanding of endogenous ligands of the receptor. Understanding the endogenous activation of TRPV1 as well as its expression, regulation, and post-signaling pathways will be necessary in order to understand and treat pathologies related to cardiovascular homeostasis and beyond.

Acknowledgments. This work was supported in part by National Institutes of Health (grants HL-57853, HL-73287, and DK-67620) and a grant from Michigan Economic Development Corporation. Dr. D. H. Wang is an Established Investigator of the American Heart Association. The authors would like to express their gratitude to fellows who have contributed to the aforementioned studies.

References

1. Montell, C., The TRP superfamily of cation channels, *Sci. STKE* re3 (2005).
2. Ramsey, S., Delling, M., and Clapham, D., An introduction to TRP channels, *Annu. Rev. Physiol.* **68**, 619–647 (2006).
3. Clapham, D.E., TRP channels as cellular sensors, *Nature* **426**, 517–524 (2003).
4. Padinjat, R., and Andrews, S., TRP channels at a glance, *J. Cell Sci.* **117**(24), 5707–5709 (2004).
5. Hellwig, N., Albrecht, N., Harteneck, C., Schultz, G., and Schaefer, M., Homo-and heteromeric assembly of TRPV channel subunits, *J. Cell Sci.* **118**, 917–928 (2005).
6. Smith, G.D., Gunthorpe, M.J., Kelsell, R.E., Hayes, P.D., Reilly, P., Facer, P., Wright, J.E., Jerman, J.C., Walhin, J.P., Ool, L., Egerton, J., Charles, K.J., Smart, D., Randall, A.D., Anand, P., and Davis, J.P., TRPV3 is a temperature-sensitive vanilloid receptor-like protein, *Nature* **418**, 186–190 (2002).
7. Liedtke, W., TRPV4 as osmosensor: a transgenic approach, *Pflügers Arch–Eur. J. Physiol.* **451**, 176–180 (2005).
8. Nijenhuis, T., Hooenderop, J.G.J., and Bindels, R.J.M., TRPV5 and TRPV6 in Ca^{2+} (re)absorption: regulating Ca^{2+} entry at the gate, *Pflügers Arch. Eur. J. Physiol.* **451**, 181–192 (2005).
9. Reid, G., ThermoTRP channels and cold sensing: what are they up to? *Pflügers Arch. Eur. J. Physiol.* **451**, 250–263 (2005).
10. Nilius, B., and Voets, T., TRP channels: a TR(I)P through a world of multifunctional cation channels, *Pflügers Arch. Eur. J. Physiol.* **451**, 1–10 (2005).
11. Clapham, D.E., Montell, C., Schultz, G., and Julius, D., The TRP ion channel family, *IUPHAR Compendium, TRP Channels* (2002).
12. Montell, C., Birnbaumer, L., and Flockerzi, V., The TRP channels, a remarkably functional family, *Cell* **108**, 595–598 (2002).
13. Desai, B.N., and Clapham, D.E., TRP channels and mice deficient in TRP channels, *Pflügers Arch. Eur. J. Physiol.* **451**, 11–18 (2005).
14. Caterina, M.J., Schumacher, M.A., Tominaga, M., Rosen, T.A., Levine, J.D., and Julius, D., The capsaicin receptor: a heat-activated ion channel in the pain pathway, *Nature* **389**, 816–824 (1997).
15. Jordt, S., McKemy, D.D., and Julius, D., Lessons from peppers and peppermint: the molecular logic of thermosensation, *Curr. Opin. Neurobio.* **13**, 487–492 (2003).
16. Agopyan, N., Head, J., Yu, S., and Simon, S.A., TRPV1 receptors mediate particulate matter-induced apoptosis, *Am. J. Physiol. Lung Cell Mol. Physiol.* **283**, L563–L572 (2004).
17. Birder, L.A., Kanai, A.J., de Groat, W.C., Kiss, S., Nealen, M.L., Burke, N.E., Dineley, K.E., Watkins, S., Reynolds, I.J., and Caterina, M.J., Altered urinary bladder function in mice lacking the vanilloid receptor TRPV1, *Nature Neurosci.* **5**(9), 856–860 (2002).

18. Lyall, V., Heck, G.L., Vinnikova, A.K., Ghosh, S., Phan, T.T., Alam, R.I. Russell, O.F., Malik, S.A., Bigbee, J.W., and DeSimone, J.A., The mammalian amiloride-insensitive non-specific salt taste receptor is a vanilloid receptor-1 variant, *J. Physiol.* **558**(1), 147–159 (2004).
19. Planells-Cases, R., Garcia-Sanz, N., Morenilla-Palao, C., and Ferrer-Montiel, A., Functional aspects and mechanisms of TRPV1 involvement in neurogenic inflammation that leads to thermal hyperalgesia, *Pflügers Arch. Eur. J. Physiol.* **451**, 151–159 (2005).
20. O'Neil, R.G., and Brown, R.C., The vanilloid receptor family of calcium-permeable channels: molecular integrators of microenvironmental stimuli, *News Physiol. Sci.* **18**, 226–231 (2003).
21. Gunthorpe, M.J., Benham, C.D., Randall, A., and Davis, J.B., The diversity in the vanilloid (TRPV) receptor family of ion channels, *Trends Pharmaco.l Sci.* **23**(4), 183–191 (2002).
22. Tominaga, M., Caterina, M.J., Malmberg, A.B., Rosen, T.A., Gilbert, H., Skinner, K., Raumann, B.E., Basbaum, A.I., and Julius, D., The cloned capsaicin receptor integrates multiple pain-producing stimuli, *Neuron* **21**, 531–543 (1998).
23. Ferrer-Montiel, A., Garcia-Martinez, C., Morenilla-Palao, C., Garcia-Sanz, N., Fernandez-Carvajal, A., Fernandez-Ballester, G., and Planells-Cases, R., Molecular architecture of the vanilloid receptor, *Eur. J. Biochem.* **271**, 1820–1826 (2004).
24. Schumacher, M.A., Moff, I., Sudanagunta, S.P., and Levine, J.D., Molecular cloning of an N-terminal splice variant of the capsaicin receptor, *J. Biol. Chem.* **275**(4), 2756–2762 (2000).
25. Tominaga, M., and Tominaga, T., Structure and function of TRPV1, *Pflügers Arch.–Eur. J. Physiol.* **451**, 143–150 (2005).
26. Vlachova, V., Teisinger, J., Susankova, K., Lyfenko, A., Ettrich, R., and Vyklicky, I., Functional role of C terminal cytoplasmic tail of rat vanilloid receptor 1, *J. Neurosci.* **23**(4), 1340–1350 (2003).
27. Numazaki, M., Tominaga, T., Takeuchi, K., Murayama, N., Toyooka, H., and Tominaga, M., Structural determinant of TRPV1 desensitization interacts with calmodulin, *Proc. Natl. Acad. Sci. USA* **100**(13), 8002–8006 (2003).
28. Kedei, N., Szabo, T., Lile, J.D., Treanor, J.J., Olah, Z., Iadarola, M.J., and Blumberg, P.M., Analysis of the native quaternary structure of vanilloid receptor 1, *J. Biol. Chem.* **276**(30), 28613–28619 (2001).
29. Kuzhikandathil, E.V., Wang, H., Szabo, T., Morozova, N., Blumberg, P.M., and Oxford, G.S., Functional analysis of capsaicin receptor (vanilloid receptor subtype 1) multimerization and agonist responsiveness using a dominant negative mutation, *J. Neurosci.* **21**(22), 8697–8706 (2001).
30. Garcia-Sanz, N., Fernandez-Carvajal, A., Morenilla-Palao, C., Planells-Cases, R., Fajardo-Sanchez, E., Fernandez-Ballester, G., and Ferrer-Montiel, A., Identification of a tetramerization domain in the C terminus of the vanilloid receptor, *J. Neurosci.* **24**(23), 5307–5314 (2004).
31. Xue, Q., Yu, Y., Trilk, S.L., Jong, B.E., and Schumacher, M.A., The genomic organization of the gene encoding the vanilloid receptor: evidence for multiple splice variants, *Genomics* **76**(1–3), 14–20 (2001).
32. Caterina, M.J., Julius, D., The vanilloid receptor: a molecular gateway to the pain pathway, *Annu. Rev. Neurosci.* **24**, 487–517 (2001).
33. Ma, Q.P., Expression of capsaicin receptor (VR1) by myelinated primary afferent neurons in rats, *Neurosci. Lett.* **319**, 87–90 (2002).

34. Kobayashi, K., Fukuoka, T., Obata, K., Yamanaka, H., Dai, Y., Tokunaga, A., and Noguchi, K., Distinct expression of TRPM8, TRPA1, and TRPV1 mRNAs in rat primary afferent neurons with A/C-fibers and co-localization Trk receptors, *J. Comp. Neurol.* **493**, 596–606 (2005).
35. Michael, G.J., and Priestley, J.V., Differential expression of the mRNA for the vanilloid receptor subtype 1 in cells of the adult rat dorsal root and nodose ganglia and its downregulation by axotomy, *J. Neurosci.* **19**(5), 1844–1854 (1999).
36. Guo, A., Vulchanova, L., Wang, J., Li, X., and Elde, R., Immunocytochemical localization of the vanilloids receptor 1 (VR1): relationship to neuropeptides, the $P2X_3$ purinoceptor and IB4 binding sites, *Eur. J. Neurosci* .**11**, 946–958 (1999).
37. Funakoshi, K., Nakano, M., Atobe, Y., Goris, R.C., Kadota, T., and Yazama, F., Differential development of TRPV1-expressing sensory nerves in peripheral organs, *Cell Tissue Res.* **323**, 27–41 (2006).
38. Szallasi, A., Vanilloid (capsaicin) receptors in health and disease, *Am. J. Clin. Pathol.* **118**, 110–121 (2002).
39. Sanchez, J.F., Krause, J.E., and Cortright, D.N., The distribution and regulation of vanilloid receptor VR1 and VR1 5' splice variant RNA expression in rat, *Neurosci.* **107**(3), 373–381 (2001).
40. Birder, L.A., Kanai, A.J., de Groat, W.C., Kiss, S., Nealen, M.L., Burke, N.E., Dineley, K.E., Watkins, S., Reynolds, I.J., and Caterina, M.J., Vanilloid receptor expression suggests a sensory role for urinary bladder epithelial cells, *Proc. Natl. Acad. Sci. USA* **98**(23), 13396–13401 (2001).
41. Avelino, A., Cruz, C., Nagy, I., and Cruz, F., Vanilloid receptor 1 expression in the rat urinary tract, *Neuroscience* **109**(4) 787–798 (2002).
42. McIntyre, P., McLatchie, L.M., Chambers, A., Phillips, E., Clarke, M., Savidge, J., Toms, C., Peacock, M., Shah, K., Winter, J., Weerasakera, N., Webb, M., Rang, H.P., Bevan, S., and James, I.F., Pharmacological differences between the human and rat vanilloid receptor 1 (VR1), *Br. J. Pharmacol.* **132**, 1084–1094 (2001).
43. Zvara, A., Bencsik, P., Fodor, G., Csont, T., Hackler, L., Jr., Dux, M., Furst, S., Jancs, G., Puskas, L.G., and Ferdinandy, P., Capsaicin-sensitive sensory neurons regulate myocardial function and gene expression pattern of rat hearts: a DNA microarray study, *FASEB J.* **20**, 160–162 (2006).
44. Nozawa, Y., Nishihara, K., Yamamoto, A., Nakano, M., Ajioka, H., and Matsuura, N., Distribution and characterization of vanilloid receptors in the rat stomach, *Neurosci. Lett.* **309**, 33–36 (2001).
45. Br, T., Maurer, M., Modarres, S., Lewin, N.E., Brodie, C., Ács, G., Ács, P., Paus, R., and Blumberg, P.M., Characterization of functional vanilloid receptors expressed by mast cells, *Blood* **91**, 1332–1340 (1998).
46. Yang, X., Lin, M., McIntosh, L.S., and Sham, J.S.K., Functional expression of transient receptor potential melastatin- (TRPM) and vanilloid-related (TRPV) channels in pulmonary arterial and aortic smooth muscle, *Am. J. Physiol. Lung Cell Mol. Physiol.* **290**(6), L1267–1276 (2005).
47. Denda, M., Fuziwara, S., Inoue, K., Denda, S., Akamatsu, H., Tomitaka, A., and Matsunaga, K., Immunoreactivity of VR1 on epidermal keratinocyte of human skin, *Biochem. Biophys. Res. Commun.* **285**, 1250–1252 (2001).
48. Inoue, K., Koizumi, S., Fuziwara, S., Denda, S., Inoue, K., and Denda, M., Functional vanilloid receptors in cultured normal human epidermal keratinocytes, *Biochem. Biophys. Res. Commun.* **291**, 124–129 (2002).

49. Ständer, S., Moormann, C., Schumacher, M., Buddenkotte, J., Artuc, M., Shpacovitch, V., Brzoska, T., Lippert, U., Henz, B.M., Luger, T.A., Metze, D., and Steinhoff, M., Expression of vanilloid receptor subtype 1 in cutaneous sensory nerve fibers, mast cells and epithelial cells of appendage structures, *Exp. Dermatol.* **13**, 129–139 (2004).
50. Apostolidis, A., Brady, C.M., Yiangou, Y., Davis, J., Fowler, C.J., and Anand, P., Capsaicin receptor TRPV1 in urothelium of neurogenic human bladders and effect of intravesical resiniferatoxin, *Urology* **65**, 400–405, (2005).
51. Cortright, D.N., Crandall, M., Sanchez, J.F., Zou, T., Krause, J.E., and White, G., The tissue distribution and functional characterization of human VR1, *Biochem. Biophys. Res. Comm.* **281**, 1183–1189 (2001).
52. Mezey, E., Toth, Z.E., Cortright, D.N., Arzubi, M.K., Krause, J.E., Elde, R., Guo, A., Blumberg, P.M., and Szallasi, A., Distribution of mRNA for vanilloid receptor subtype 1 (VR1), and VR1-like immunoreactivity, in the central nervous system of the rat and human, *Proc. Natl. Acad. Sci. USA* **97**, 3655–3660 (2000).
53. Nagy, I., Santha, P., Jancso, G., and Urban, L., The role of the vanilloid (capsaicin) receptor (TRPV1) in physiology and pathology, *Eur. J. Pharmacol.* **500**, 351–369 (2004).
54. Caterina, M.J., Vanilloid receptors take a TRP beyond the sensory afferent, *Pain* **105**, 5–9 (2003).
55. Ross, R.A., Anandamide and vanilloid TRPV1 receptors, *Br. J. Pharmacol.* **140**, 790–801 (2003).
56. Zygmunt, P.M., Petersson, J., Andersson, D.A., Chuang, H., Sorgard, M., Di Marzo, V., Julius, D., and Hogestatt, E.D., Vanilloid receptors on sensory nerves mediate the vasodilator action of anandamide, *Nature* **400**, 452–456 (1999).
57. Smart, D., Gunthorpe, M.J., Jerman, J.C., Nasir, S., Gray, J., Muir, A.I., Chambers, J.K., Randall, A.D., and Davis, J.B., The endogenous lipid anandamide is a full agonist at the human vanilloid receptor (hVR1), *Br. J. Pharmacol.* **129**, 227–230 (2000).
58. Olah, Z., Karai, L., and Iadarola, M.J., Anandamide activates vanilloid receptor 1 (VR1) at acidic pH in dorsal root ganglia neurons and cells ectopically expressing VR1, *J. Biol. Chem.* **276**(33), 31163–31170 (2001).
59. Van der Stelt, M., Trevisani, M., Vellani, V., de Petrocellis, L., Moriello, A.S., Campi, B., McNaughton, P., Geppetti, P., and Di Marzo, V., Anandamide acts as an intracellular messenger amplifying Ca^{2+} influx via TRPV1 channels, *Eur. Mol. Bio. Org. J.* **24**, 3026–3037 (2005).
60. Malinowska, B., Kwolek, G., and Gothert, M., Anandamide and methanandamide induce both vanilloid VR1- and cannabinoid CB_1 receptor-mediated changes in heart rate and blood pressure in anaesthetized rats, *Naunyn-Schmiedeberg's Arch. Pharmacol.* **364**, 562–569 (2001).
61. Ralevic, V., Kendall, D.A., Randall, M.D., Zygmunt, P.M., Movahed, P., and Hogestatt, E.D., Vanilloid receptors on capsaicin-sensitive sensory nerves mediate relaxation to methanandamide in the rat isolated mesenteric arterial bed and small mesenteric arteries, *Br. J. Pharmacol.* **130**, 1483–1488 (2000).
62. Huang, S.M., and Walker, J.M., Enhancement of spontaneous and heat-evoked activity in spinal nociceptive neurons by the endovanilloid/endocannabinoid N-arachidonyldopamine (NADA), *J. Neurophysiol.* **95**, 1207–1212 (2006).
63. Huang, S.M., Bisogno, T., Trevisani, M., Al-Hayani, A., De Petrocellis, L., Fezza, F., Tognetto, M., Petros, T.J., Krey, J.F., Chu, C.J., Miller, J.D., Davies, S.N., Geppetti,

P., Walker, J.M., and Di Marzo, V., An endogenous capsaicin-like substance with high potency at recombinant and native vanilloid VR1 receptors, *Proc. Natl. Acad. Sci. USA* **99**(12), 8400–8405 (2002).
64. Appendino, G., De Petrocellis, L., Trevisani, M., Minassi, A., Daddario, N., Moriello, A.S., Gazzieri, D., Ligresti, A., Campi, B., Fontana, G., Pinna, C., Geppetti, P., and Di Marzo, V., Development of the first ultra-potent "capsaicinoid" agonist at transient receptor potential vanilloid type 1 (TRPV1) channels and its therapeutic potential, *J. Pharmacol. Exp. Ther.* **312**(2), 561–570 (2005).
65. Hu, H., Gu, Q., Wang, C., Colton, C.K., Tang, J., Kinoshita-Kawada, M., Lee, L., Wood, J.D., and Zhu, M.X., 2-Aminoethoxydiphenyl borate is a common activator of TRPV1, TRPV2, and TRPV3, *J. Biol. Chem.* **279**(34), 35741–35748 (2004).
66. Gu, Q., Lin, R., Hu, H., Zhu, M.X., and Lee, L., 2-aminoethoxydiphenyl borate stimulates pulmonary C neurons via the activation of TRPV channels, *AJP–Lung* **288**, 932–941 (2005).
67. Ahern, G.P., Activation of TRPV1 by the satiety factor oleoylethanolamide, *J. Biol. Chem.* **278**(33), 30429–30434 (2003).
68. Chu, C.J., Huang, S.M., De Petrocellis, L., Bisogno, T., Ewing, S.A., Miller, J.D., Zipkin, R.E., Daddario, N., Appendino, G., Di Marzo, V., and Walker, J.M., N-oleoyldopamine, a novel endogenous capsaicin-like lipid that produces hyperalgesia, *J. Biol. Chem.* **278**(16), 13633–13639 (2003).
69. Trevisani, M., Smart, D., Gunthorpe, M.J., Tognetto, M., Barbieri, M., Campi, B., Amadesi, S., Gray, J., Jerman, J.C., Brough, S.J., Owen, D., Smith, G.D., Randall, A.D., Harrison, S., Bianchi, A., Davis, J.B., and Geppetti, P., Ethanol elicits and potentiates nociceptor responses via the vanilloid receptor-1, *Nature Neurosci.* **5**(6), 546–551 (2002).
70. Gazzieri, D., Trevisani, M., Tarantini, F., Bechi, P., Masotti, G., Gensini, G.F., Castellani, S., Marchionni, N., Geppetti, P., and Harrison, S., Ethanol dilates coronary arteries and increases coronary flow via transient receptor potential vanilloid 1 and calcitonin gene-related peptide, *Cardiovasc. Res.* **70**, 589–599 (2006).
71. Trevisani, M., Smart, D., Gunthorpe, M.J., Tognetto, M., Barbieri, M., Campi, B., Amadesi, S., Gray, J., Jerman, J.C., Brough, S.J., Owen, D., Smith, G.D., Randall, A.D., Harrison, S., Bianchi, A., Davis, J.B., and Geppetti, P., Ethanol elicits and potentiates nociceptor responses via the vanilloid receptor-1, *Nature Neurosci.* **5**(6), 546–551 (2002).
72. Hwang, S.W., Cho, H., Kwak, J., Lee, S., Kang, C., Jung, J., Cho, S., Min, K.H., Suh, Y., Kim, D., and Oh, U., Direct activation of capsaicin receptors by products of lipoxygenases: Endogenous capsaicin-like substances, *Proc. Natl. Acad. Sci. USA* **97**(11), 6155–6160 (2000).
73. Scotland, R.S., Chauhan, S., Davis, C., De Felipe, C., Hunt, S., Kabir, J., Kotsonis, P., Oh, U., and Ahluwalia, A., Vanilloid receptor TRPV1, sensory C-fibers, and vascular autoregulation, *Circ. Res.* **95**, 1027–1034 (2004).
74. Nagy, I., and Rang, H., Noxious heat activates all capsaicin-sensitive and also a subpopulation of capsaicin-insensitive dorsal root ganglion neurons, *Neurosci.* **88**(4), 995–997 (1999).
75. Welch, J.M., Simon, S.A., and Reinhart, P.H., The activation mechanism of rat vanilloid receptor 1 by capsaicin involves the pore domain and differs from the activation by either acid or heat, *Proc. Natl. Acad. Sci. USA* **97**(25), 13889–13894 (2000).

76. Tominaga, M., Wada, M., and Masu, M., Potentiation of capsaicin receptor activity by metabotropic ATP receptors as a possible mechanism for ATP-evoked pain and hyperalgesia, *Proc. Natl. Acad. Sci. USA* **98**(12), 6951–6956 (2001).
77. Kikuno, S., Taguchi, K., Iwamoto, N., Yamano, S., Cho, A.K., Froines, J.R., and Kumagai, Y., 1,2-Naphthoquinone activates vanilloid receptor 1 through increased protein tyrosine phosphorylation, leading to contraction of guinea pig trachea, *Tox. Appl. Pharmacol.* **210**, 47–54 (2006).
78. Wardle, K.A., Ranson, J., and Sanger, G.J., Pharmacological characterization of the vanilloid receptor in the rat dorsal spine cord, *Br. J. Pharmacol.* **121**, 1012–1016 (1997).
79. Rigoni, M., Trevisani, M., Gazzieri, D., Nadaletto, R., Tognetto, M., Creminon, C., Davis, J.B., Campi, B., Amadesi, S., Geppetti, P., and Harrison, S., Neurogenic responses mediated by vanilloid receptor-1 (TRPV1) are blocked by the high affinity antagonist, iodo-resiniferatoxin, *Br. J. Pharmacol.* **138**, 977–985 (2003).
80. Xi, N., Bo, Y., Doherty, E.M., Fotsch, C., Gavva, N.R., Han, N., Hungate, R.W., Klionsky, L., Liu, Q., Tamir, R., Xu, S., Treanor, J.J.S., and Norman, M.H., Synthesis and evaluation of thiazole carboxamides as vanilloid receptor 1 (TRPV1) antagonists, *Bioorg. Med. Chem. Ltrs.* **15**, 5211–5217 (2005).
81. Kouhen, R.E., Surowy, C.S., Bianchi, B.R., Neelands, T.R., McDonald, H.A., Niforatos, W., Gomtsyan, A., Lee, C., Honore, P., Sullivan, J.P., Jarvis, M.F., and Faltynek, C.R., A-425619 {1-Isoquinolin-5-yl-3-(4-trifluoromethyl-benzyl)-urea}, a novel and selective transient receptor potential type V1 receptor antagonist, blocks channel activation by vanilloids, heat and acid, *J. Pharmacol. Exp. Therap.* **314**(1), 400–409 (2005).
82. McGaraughty, S., Chu, K.L., Faltynek, C.R., and Jarvis, M.F., Systemic and site-specific effects of A-425619, a selective TRPV1 receptor antagonist, on wide dynamic range neurons in CFA-treated and uninjured rats, *J. Neurophysiol.* **95**, 18–25 (2006).
83. Xi, N., Bo, Y., Doherty, E.M., Fotsch, C., Gavva, N.R., Han, N., Hungate, R.W., Klionsky, L., Liu, Q., Tamir, R., Xu, S., Treanor, J.J.S., and Norman, M.H., Synthesis and evaluation of thiazole carboxamides as vanilloid receptor 1 (TRPV1) antagonists, *Bioorg. Med. Chem. Ltr.s* **15**, 5211–5217 (2005).
84. Marquez, N., De Petrocellis, L., Caballero, F.J., Macho, A., Schiano-Moriello, A., Minassi, A., Appendino, G., Munoz, E., and Di Marzo, V., Iodinated N-acylvanillamines: potential "multiple-target" anti-inflammatory agents acting via the inhibition of t-cell activation and antagonism at vanilloid TRPV1 channels, *Mol. Pharmacol.* **69**(4), 1373–1382 (2006).
85. Szallasi, A., and Appendino, G., Vanilloid receptor TRPV1 antagonists as the next generation of painkillers. Are we putting the cart before the horse? *J. Med. Chem.* **47**(11), 2717–2723 (2004).
86. Jung, J., Lee, S., Hwang, S.W., Cho, H., Shin, J., Kang, Y., Kim, S., and Oh, U., Agonist recognition sites in the cytosolic tails of vanilloid receptor 1, *J. Biol. Chem.* **277**(46), 44448–44454 (2002).
87. Jordt, S., Tominaga, M., and Julius, D., Acid potentiation of the capsaicin receptor determined by a key extracellular site, *Proc. Natl. Acad. Sci. USA* **97**(14), 8134–8139 (2000).
88. Gavva, N.R., Klionsky, L., Qu, Y., Shi, L., Tamir, R., Edenson, S., Zhang, T.J., Viswanadhan, V.N., Toth, A., Pearce, L.V., Vanderah, T.W., Porreca, F., Blumberg, P.M., Lile, J., Sun, Y., Wild, K., Louis, J., and Treanor, J.J.S., Molecular determinants of vanilloid sensitivity in TRPV1, *J. Biol. Chem.* **279**(19), 20283–20295 (2004).

89. Jordt, S., Julius, D., Molecular basis for species-specific sensitivity to "hot" chili peppers, *Cell* **108**, 421–430 (2002).
90. Sutton, K.G., Garrett, E.M., Rutter, A.R., Bonnert, T.P., Jarolimek, W., and Seabrook, G.R., Functional characterization of the S512Y mutant vanilloid human TRPV1 receptor, *Br. J. Pharm.* **146**, 702–711 (2005).
91. Jhaveri, M.D., Elmes, S.J., Kendall, D.A., and Chapman, V., Inhibition of peripheral vanilloid TRPV1 receptors reduces noxious heat-evoked responses of dorsal horn neurons in naïve, carrageenan-inflamed and neuropathic rats, *Eur. J. Neurosci.* **22**(2), 361–70 (2005).
92. Premkumar, L.S., and Ahern, G.P., Induction of vanilloid receptor channel activity by protein kinase C, *Nature* **408**, 985–990 (2000).
93. Mandadi, S., Numazaki, M., Tominaga, M., Bhat, M.B., Armati, P.J., and Roufogalis, B.D., Activation of protein kinase C reverses capsaicin-induced calcium-dependent desensitization of TRPV1 ion channels, *Cell Calcium* **35**, 471–478 (2004).
94. Numazaki, M., Tominaga, T., Toyooka, H., and Tominaga, M., Direct phosphorylation of capsaicin receptor VR1 by protein kinase C and identification of two target serine residues, *J. Biol. Chem.* **277**(16), 13375–13378 (2002).
95. Bhave, G., Ju, H., Glauner, K.S., Zhu, W., Wang, H., Brasier, D.J., Oxford, G.S., and Gereau, R.W., IV, Protein kinase C phosphorylation sensitizes but does not activate the capsaicin receptor transient receptor potential vanilloid 1 (TRPV1), *Proc. Natl. Acad. Sci. USA* **100**(21), 12480–12485 (2003).
96. Liu, B., Ma, W., Ryu, S., and Qin, F., Inhibitory modulation of distal C-terminal on protein kinase C-dependent phosphor-regulation of rat TRPV1 receptors, *J. Physiol.* **560**(3), 627–638 (2004).
97. Chuang, H., Prescott, E.D., Kong, H., Shields, S., Jordt, S., Basbaum, A.I., Chao, M.V., and Julius, D., Bradykinin and nerve growth factor release the capsaicin receptor from PtdIns(4, 5)P_2-mediated inhibition, *Nature* **411**, 957–962 (2001).
98. Prescott, E.D., and Julius, D., A modular PIP_2 binding site as a determinant of capsaicin receptor sensitivity, *Science* **300**, 1284–1288 (2003).
99. Morenilla-Palao, C., Planells-Cases, R., Garcia-Sanz, N., and Ferrer-Montiel, A., Regulated exocytosis contributes to protein kinase C potentiation of vanilloid receptor activity, *J. Biol. Chem.* **279**(24), 25665–25672 (2004).
100. Sugiura, T., Tominaga, M., Katsuya, H., and Mizumura, K., Bradykinin lowers the threshold temperature for heat activation of vanilloid receptor 1, *J. Neurophysiol.* **88**, 544–548 (2002).
101. Tang, H., Inoue, A., Oshita, K., and Nakata, Y., Sensitization of vanilloid receptor 1 induced by bradykinin via the activation of second messenger signaling cascades in rat primary afferent neurons, *Eur. J. Pharmacol.* **498**, 37–43 (2004).
102. Obreja, O., Rathee, P.K., Lips, K.S., Distler, C., and Kress, M., IL-1ß potentiates heat-activated currents in rat sensory neurons: involvement of IL-1RI, tyrosine kinase, and protein kinase C, *FASEB J.* **16**, 1497–1503 (2002).
103. Lopshire, J.C., and Nicol, G.D., The cAMP transduction cascade mediates the prostaglandin E_2 enhancement of the capsaicin-elicited current in rat sensory neurons: whole-cell and single-channel studies, *J. Neurosci.* **18**(16), 6081–6092 (1998).
104. Mohapatra, D.P., and Nau, C., Desensitization of capsaicin-activated currents in the vanilloid receptor TRPV1 is decreased by the cyclic AMP-dependent protein kinase pathway, *J. Biol. Chem.* **278**(50), 50080–50090 (2003).

105. Rathee, P.K., Distler, C., Obreja, O., Neuhuber, W., Wang, G.K., Wang, S.Y., Nau, C., and Kress, M., PKA/AKAP/VR-1 module: A common link of G_s-mediated signaling to thermal hyperalgesia, *J. Neurosci.* **22**(11), 4740–4745.
106. Docherty, R.J., Yeats, J.C., Bevan, S., and Boddeke, H.W.G.M., Inhibition of calcineurin inhibits the desensitization of capsaicin-evoked currents in cultured dorsal root ganglion neurons from adult rats, *Flügers Arch.–Eur. J. Physiol.* **431**, 828–837 (1996).
107. Mohapatra, D.P., and Nau, C., Regulation of Ca^{2+} -dependent desensitization in the vanilloid receptor TRPV1 by calcineurin and cAMP-dependent protein kinase, *J. Biol. Chem.* **280**(14), 13424–13432 (2005).
108. Jung, J., Shin, J.S., Lee, S., Hwang, S.W., Koo, J., Cho, H., and Oh, U., Phosphorylation of vanilloid receptor 1 by Ca^{2+}/calmodulin-dependent kinase II regulates its vanilloid binding, *J. Biol. Chem.* **279**(8), 7048–7054 (2004).
109. Wang, C., Hu, H., Colton, C.K., Wood, J.D., and Zhu, M.X., An alternative splicing product of the murine trpv1 gene dominant negatively modulates the activity of Trpv1 channels, *J. Biol. Chem.* **279**(36), 37423–37430 (2004).
110. Tian, W., Fu, Y., Wang, D.H., and Cohen, D.M., Regulation of TRPV1 by a novel renally expressed rat TRPV1 splice variant, *AJP-Renal* **290**, 117–126 (2006).
111. Rosenbaum, T., Gordon-Shaag, A., Munari, M., and Gordon, S.E., Ca^{2+}/Calmodulin modulates TRPV1 activation by capsaicin, *J. Gen. Physiol.* **123**, 53–62 (2004).
112. Amaya, F., Oh-hashi, K., Naruse, Y., Iijima, N., Ueda, M., Shimosato, G., Tominaga, M., Tanaka, Y., and Tanaka, M., Local inflammation increases vanilloid receptor 1 expression within distinct subgroups of DRG neurons, *Brain Res.* **963**(1-2), 190–196 (2003).
113. Kim, S., Kang, C., Shin, C.Y., Hwang, S.W., Yang, Y.D., Shim, W.S., Park, M., Kim, E., Kim, M., Kim, B., Cho, H., Shin, Y., and Oh, U., TRPV1 recapitulates native capsaicin receptor in sensory neurons in association with Fas-associated factor 1, *J. Neurosci.* **26**(9), 2403–2412 (2006).
114. Van Buren, J.J., Bhat, S., Rotello, R., Pauza, M.E., and Premkumar, L.S., Sensitization and translocation of TRPV1 by insulin and IGF-1, *Mol. Pain* **1**, 17 (2005).
115. Vyklicky, L., Lyfenko, A., Susankova, K., Teisinger, J., and Vlachova, V., Reducing agent dithiothreitol facilitates activity of the capsaicin receptor VR-1, *Neuroscience* **111**(3), 435–441 (2002).
116. Wang, D.H., and Li, J., Antihypertensive mechanisms underlying a novel salt-sensitive hypertensive model induced by sensory denervation, *Hypertension* **33**(15), 499–503 (1999).
117. Supowit, S.C., Ethridge, R.T., Zhao, H., Katki, K.A., and DiPette, D.J., Calcitonin gene-related peptide and substance P contribute to reduced blood pressure in sympathectomized rats, *AJP-Heart* **289**, 1169–1175 (2005).
118. van der Stelt, M., and Di Marzo, V., Endovanilloids. Putative endogenous ligands of transient receptor potential vanilloid 1 channels, *Eur. J. Biochem.* **271**, 1827–1834 (2004).
119. Naeini, R.S., Witty, M., Seguela, P., and Bourque, C.W., An N-terminal variant of Trpv1 channel is required for osmosensory transduction, *Nature Neurosci.* **9**(1), 93–98 (2006).
120. Oliet, S.H., and Bourque, C.W., Mechanosensitive channels transducer osmosensitivity in supraoptic neurons, *Nature* **364**, 341–343 (1993).

121. Winter, J., Dray, A., Wood, J.N., Yeats, J.C., and Bevan, S., Cellular mechanism of action of resiniferatoxin: a potent sensory neuron excitotoxin, *Brain Res.* **520**, 13–40 (1990).
122. Bevan, S., and Forbes, C.A., Membrane effects of capsaicin on rat dorsal root ganglion neurons in culture, *J. Physiol.* **398**, 28 (1988).
123. Forbes, C.A., and Bevan, S., Single channels activated by capsaicin in patches of membrane from adult rat sensory neurons in culture, *Neurosci. Lett.* **S3**, 32 (1998).
124. Marsh, S.J., Stansfeld, C.E., Brown, D.A., Davey, R., and McCarthy, D., The mechanism of action of capsaicin on sensory C-type neurons and their axons *in vitro*, *Neuroscience*, **23**, 275–289 (1987).
125. Wood, J.N., Winter, J., James, I.F., Rang, H.P., Yeats, J., and Bevan, S., Capsaicin-induced ion fluxes in dorsal root ganglion cells in culture, *J. Neurosci.***8**, 3208–3220 (1988).
126. Maggi, C.A., The pharmacological modulation of neurotransmitter release, in: *Capsaicin in the Study of Pain,* edited by J. Wood. Harcourt Brace & Company, New York, 1993, pp. 161–189.
127. Wimalawansa, S.J., Calcitonin gene-related peptide and its receptors: molecular genetics, physiology, pathophysiology, and therapeutic potentials, *Endocr. Rev.* **17**, 533–585 (1996).
128. Arendshorst, W.J., Cook, M.A., and Mills, I.H., Effect of substance P on proximal tubular reabsorption in the rat, *Am. J. Physiol.* **230**, 1662–1667 (1976).
129. Shekhar, Y.C., Anand, I.S., Sarma, R., Ferrari, R., Wahi, P.L., and Poole-Wilson, P.A., Effects of prolonged infusion of human alpha calcitonin gene-related peptide on hemodynamics, renal blood flow and hormone levels in congestive heart failure, *Am. J. Cardiol.* **67**, 732–736 (1991).
130. Breimer, L.H., MacIntyre, I., and Zaidi, M., Peptides from the calcitonin genes: molecular genetics, structure, and function, *Biochem. J.* **255**, 377–390 (1988).
131. Preibisz, J.J., CGRP and regulation of human cardiovascular homeostasis, *Am. J. Hypertens.* **6**, 434–450 (1993).
132. Zaidi, M., Moonga, B.S., Bevis, P.J., Bascal, Z.A., and Breimer, L.H., The calcitonin gene peptides: biology and clinical relevance, *Crit. Rev. Clin. Lab. Sci.* **28**, 109 (1999).
133. McEwan, J., Legon, S., Wimalawansa, S.J., Zaidi, M., Dollery, C.T., and MacIntyre, I., Calcitonin gene-related peptide: a review of its biology and relevance to the cardiovascular system, in: *Endocrine Mechanisms in Hypertension,* edited by J.H. Laragh, B.N. Brenner, and N.M. Kaplan. Raven Press, New York, 1989, p. 287.
134. Szallasi, A., and Blumberg, P.M., Vanilloid (capsaicin) receptors and mechanisms, *Pharmacol. Rev.* **51**, 159–210 (1999).
135. Wang, D.H., Wu, W., and Lookingland, K.J., Degeneration of capsaicin-sensitive sensory nerves leads to increased salt sensitivity through enhancement of sympathoexcitatory response, *Hypertension* **37**, 440–443 (2001).
136. Wang, D.H., Li, J.P., and Qiu, J.X., Salt sensitive hypertension induced by sensory denervation: Introduction of a new model. *Hypertension (Rapid Communication)* **32**(4), 649–653 (1998).
137. Wang, D.H., and Zhao, Y.Z., Increased salt sensitivity induced by impairment of sensory nerves: is nephropathy the cause? *J. Hypertens.* **21**(2), 403–409 (2003).
138. Li, J.P., and Wang, D.H., High salt induced-increase in blood pressure: Role of capsaicin-sensitive sensory nerves. *J. Hypertens.* **21**(3), 577–582 (2003).

139. Zhu, Y., Wang, Y., and Wang, D.H., Diuresis and natriuresis caused by activation of VR1-positive sensory nerves in renal pelvis of rats, *Hypertension* **46**(part 2), 992–997 (2005).
140. Li, J., and Wang, D.H., Function and regulation of the vanilloid receptor in rats fed a high salt diet, *J. Hypertens* **21**, 1525–1530 (2003).
141. Wang, Y., Kaminski, N.E., and Wang, D.H., VR1-mediated depressor effects during high salt intake.: role of anandamide, *Hypertension* **46**(part 2), 986–991 (2005).
142. Li, J.P., Kaminski, N.E., and Wang, D.H., Anandamide-induced depressor effect in spontaneously hypertensive rats: role of the vanilloid receptor (VR1). *Hypertension* **41**(2), 757–762 (2003).
143. Wang, Y., and Wang, D.H., A novel mechanixm contributing to development of dahl salt-sensitive hypertension. Role of the transient receptor potential vanilloid type 1, *Hypertension* **47** (part 2), 609–614 (2006).
144. Huang, Y., and Wang, D.H., Role of AT1 and AT2 receptor subtypes in salt sensitive hypertension induced by sensory denervation. *J. Hypertension* **19**(10), 1841–1846 (2001).
145. Huang, Y., and Wang, D.H., Role of the renin-angiotensin-aldosterone system in salt sensitive hypertension induced by sensory denervation. *Am. J. Physiol. (Heart & Circulatory Physiology)* **281** (5), H2143-H2149 (2001).
146. Ye, D., and Wang, D.H., Function and regulation of endothelin receptor subtypes in salt sensitive hypertension induced by sensory nerve degeneration. *Hypertension* **39** (2 pt 2), 673–678 (2002).
147. Wang, Y.P., Chen, A.F., and Wang, D.H., Role of ET_A receptor in salt sensitive hypertension induced by sensory nerve degeneration. *Am. J. Physiol. (Heart & Circulatory Physiology)* **289**, H2005–H2011 (2005).
148. Song, W.Z., Chen, A., and Wang, D.H., Increased salt sensitivity induced by impairment of sensory nerves: role of superoxide. *Acta Pharmacoogica. Sinica* **25**(12), 1626–1632 (2004).
149. Wang, D.H., The vanilloid receptor and hypertension, *Acta Pharmacologica Sinica* **26**(3), 286–294 (2005).
150. Nilius, B., Voets, T., and Peters, J., TRP channels in disease, *Sci. STKE* **2005**(295), re8, 2 (2005).
151. Deng, P., and Li, Y., Calcitonin gene-related peptide and hypertension, *Peptides* **26**, 1676–1685 (2005).
152. Wang, L.H., and Wang, D.H., Vanilloid receptor gene knockout impairs postischemic recovery in isolated perfused heart in mice. *Circulation* **112**(23), 3617–3623 (2005).

7
Functional Interaction Between ATP and TRPV1 Receptors

Makoto Tominaga* and Tomoko Moriyama

7.1. Introduction

Noxious thermal, mechanical, or chemical stimuli evoke pain through excitation of the peripheral terminals called nociceptors.[1,2] Many kinds of ionotropic and metabotropic receptors are involved in this process.[3–6] TRPV1, a capsaicin receptor, is a nociceptor-specific ion channel that serves as the molecular target of capsaicin, having six transmembrane domains with a short hydrophobic stretch between the fifth and sixth transmembrane domains.[7] TRPV1 can be activated not only by capsaicin but also by noxious heat (with a thermal threshold >43°C) or protons (acidification), all of which are known to cause pain *in vivo*.[8,9] The studies using TRPV1-deficient mice have shown that TRPV1 is essential for selective modalities of pain sensation and for thermal hyperalgesia induced by tissue injury.[10,11] Inflammatory pain is initiated by tissue damage/inflammation and is characterized by hypersensitivity both at the site of damage and in adjacent tissue. One mechanism underlying these phenomena is the modulation (sensitization) of ion channels such as TRPV1 that detect noxious stimuli at the nociceptor terminal. Sensitization is triggered by extracellular inflammatory mediators, such as adenosine 5′-triphosphate (ATP), bradykinin and prostaglandins. In this chapter, regulation mechanisms of TRPV1 function by the inflammatory mediators are discussed by focusing on ATP action.

7.2. ATP Release from the Cells and Its Targets

ATP is well recognized as an energy source and modulator of intracellular function, operating ubiquitously within the body. Extracellular ATP is also a well-recognized autocrine and paracrine regulator of a multitude of physiological functions at cellular as well as at organ levels.[12,13] To function as an extracellular regulator, ATP has to be released from the cells. Mechanical stress and osmotic swelling are the

*Section of Cell Signaling, Okazaki Institute for Integrative Bioscience, National Institutes of Natural Sciences, Okazaki 444-8787, Japan; E-mail: tominaga@nips.ac.jp

most effective physiological stimuli for ATP release.[12,14] When tissues are damaged, ATP is released into the extracellular spaces from surrounding damaged or inflamed tissues. ATP is also released by exocytosis as a co-transmitter with norepinephrine and neuropeptide Y from sympathetic nerve terminal varicosities. Within the heart, it is known that ATP is released into the interstitial spaces in response to a variety of physiological and pathophysiological stimuli, including hypoxia, ischemia or ischemic preconditioning, mechanical stretching and stimulation by catecholamines.[15] For example, brief hypoxia was reported to induce ATP release from isolated rat ventricular myocytes.[16] In the whole heart, ATP from other sources such as microvascular endothelial cells during hyperaemia may also contribute to the ATP release.[17]

Among the various physiological effects of extracellular ATP, ATP is known to excite the nociceptive endings of nearby sensory nerves evoking a sensation of pain. In these neurons, the most widely studied targets of extracellular ATP have been ionotropic ATP (P2X) receptors.[18,19] Indeed, several P2X receptor subtypes have been identified in sensory neurons, including one ($P2X_3$) whose expression is largely confined to these cells.[20,21] It has been well documented that signaling by ATP through sensory neuron-specific $P2X_3$ and $P2X_{2/3}$ receptors mediates an acute nociceptive pathway contributing to the pain of tissue damage. Our understanding of purinergic contributions to pain sensation may be incomplete, however, given that the potential involvement of widely distributed metabotropic ATP (P2Y) receptors has not yet been widely investigated.

7.3. ATP Potentiates TRPV1 Activity in HEK 293 Cells Expressing TRPV1

Extracellular ATP has been reported to potentiate capsaicin- or proton-evoked currents through $P2Y_1$ receptors in a PKC-dependent manner in HEK293 cells expressing TRPV1.[22] Extracellular ATP was found to shift the capsaicin- or proton-dose-dependent curve to the left without changing the maximal response. In addition, ATP lowers the temperature threshold for heat activation of TRPV1 from 42°C (a temperature causing pain in our body) to 35°C, such that normally non-painful thermal stimuli such as normal body temperature are capable of activating TRPV1, thereby possibly causing pain sensation. A PKC-mediated mechanism has been thought to be involved in the sensitization of TRPV1 by ATP, based on the observation that several different PKC inhibitors blocked the ATP-induced potentiation of TRPV1 activity. Then, direct phosphorylation of TRPV1 by PKC was proven using a biochemical approach, and two target serine residues (S502 and S800) were identified.[23] In the mutant where the two serine residues were replaced with alanine, sensitization or potentiation of TRPV1 activity induced by any of three different stimuli (capsaicin, proton, or heat) upon PKC activation was almost completely abolished, suggesting the involvement of the two residues in the regulation of TRPV1 activity by PKC.

7.4. Functional Interaction Between ATP and TRPV1 at an Animal Level

In the study using TRPV1-deficient mice, ATP was proven to interact with TRPV1 at an animal level.[24] Wild-type mice exhibited ATP-induced thermal hyperalgesia upon intraplantar injection of ATP into one of their hind paws (reduction in paw-withdrawal latency to radiant heat), while the ATP-induced thermal hyperalgesia was not observed in TRPV1-deficient mice. Pharmacological analysis of ATP-induced potentiation of TRPV1 responses evoked by capsaicin in HEK293 cells expressing TRPV1 suggested the involvement of $P2Y_1$ receptors. To confirm the involvement of $P2Y_1$ receptors, behavioral analysis using $P2Y_1$-deficient mice was performed. Surprisingly, intraplantar injection of ATP produced thermal hyperalgesia in $P2Y_1$-deficient mice similar to that observed in wild-type mice, indicating that $P2Y_1$ receptors are not involved in the functional interaction between ATP and TRPV1. Electrophysiological and pharmacological analyses in mouse DRG neurons suggested the involvement of $P2Y_2$ receptors in the ATP-induced potentiation of TRPV1 activity. The results that $P2Y_2$ but not $P2Y_1$ mRNA was co-localized with TRPV1 mRNA in small-diameter DRG neurons and that UTP, a relatively selective agonist of $P2Y_2$ and $P2Y_4$ receptors, produced thermal hyperalgesia with time course similar to that observed in ATP injection, support the conclusion.

$P2Y_2$ receptors confer responsiveness to UTP and ATP to a similar extent, suggesting a possible role for UTP as an important component of pro-algesic response in the context of tissue injury. UTP has been reported to be released from ruptured cells.[25,26] In addition to non-neuronal cells, ATP can be released from a subset of small primary afferent nerves in response to capsaicin,[27] suggesting possible autocrine and paracrine mechanisms for the exacerbation of pain. Similar context can be imagined for UTP action in tissue damage. Therefore, UTP as well as ATP should be taken into account when purinergic contributions through P2Y receptors to pain sensation are examined.

ATP and UTP have recently been reported to cause CREB phosphorylation, which is likely to activate gene expression, through $P2Y_2$ receptor activation.[28] $P2Y_2$ receptor activation could increase intracellular Ca^{2+} levels, leading to various outcomes including CREB phosphorylation, activation of phospholipase A_2 and the consequent liberation of arachidonic acid, the rate-limiting step in prostaglandin.[28,29] This Ca^{2+} mobilizing process could occur more slowly than the effects on TRPV1. Both fast and slow signals downstream of $P2Y_2$ receptor activation might exist in native cells and contribute coordinately to hyperalgesia. Most attention in the pain field has focused on the role of ionotropic ATP receptors in ATP-induced nociception. The above findings together with previous studies suggest that $P2Y_2$ is also involved in this process and that ATP could act on both ionotropic and metabotropic receptors to initiate and maintain the nociceptive responses in concert.

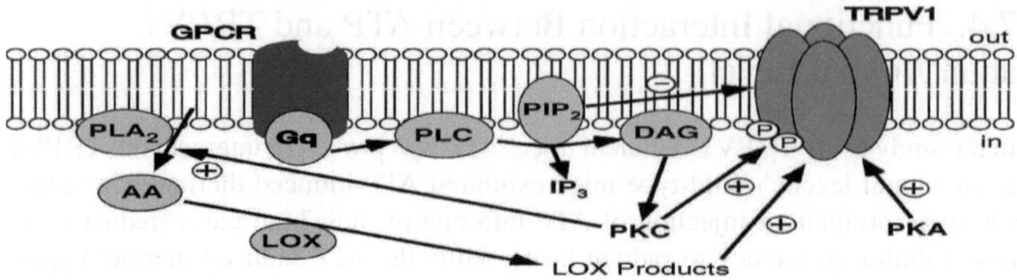

FIGURE 7.1. Proposed mechanisms involved in G_q-coupled receptor (GPCR)-mediated potentiation of TRPV1 activity. G_q-coupled receptor activation leads to production of inositol 1,4,5-trisphosphate (IP_3) and diacyl glycerol (DAG) from phosphatidylinositol-4,5-bisphosphate (PIP_2) through phospholipase C (PLC). PKC activation by DAG causes phosphorylation of TRPV1 (P), leading to functional potentiation (+). PKA also potentiates TRPV1 activity. PIP_2 inhibits TRPV1 (−). PKC activates phospholipase A_2 (PLA_2) leading to the arachidonic acid (AA) production, followed by generation of lipoxygenase (LOX)-derived products.

7.5. Involvement of Similar PKC-Dependent Mechanism and the Mechanism Downstream of PLC Activation in TRPV1 Sensitization

The functional interaction between TRPV1 and $P2Y_2$ receptors represents a novel mechanism through which extracellular ATP and UTP might cause pain in the context of inflammation and a significant interaction between an ion channel (TRPV1) and a metabotropic receptor ($P2Y_2$) to induce inflammatory pain sensation. One major consequence of $P2Y_2$ receptor stimulation is activiation of phospholipase C (PLC) through the G protein, G_q, leading to the production of inositol 1,4,5-triphosphate (IP_3) and diacylglycerol (DAG), followed by PKC activation.[19] If G_q activation underlies the potentiation of TRPV1 by extracellular ATP, stimulation of other G_q-coupled receptors may produce similar effects (Fig. 7.1). Indeed, bradykinin, a nonapeptide released into inflamed tissues, and known to induce pain and hyperalgesia to heat[30] was found to potentiate or sensitize TRPV1 through the activation of bradykinin B2 in a PKC-dependent manner.[31] Moreover, proteinase-activated receptor 2 (PAR2), known to be involved in inflammation[32] was also reported to make a functional complex with TRPV1 and may cause pain sensation in response to trypsin or tryptase released in the inflammation.[33,34] Other substances released in the inflammation that activate G_q-coupled receptors may be able to sensitize TRPV1 in a similar way.

There seems to be PKC-dependent and -independent mechanisms in the regulation of TRPV1 activity downstream of PLC activation (Fig. 7.2). Disruption of interaction between phosphatidylinositol-4,5-bisphosphate (PIP_2) and TRPV1 has also been reported to be involved in the sensitization of TRPV1 in the downstream of PLC activation. The direct PLC-mediated PIP_2 hydrolysis

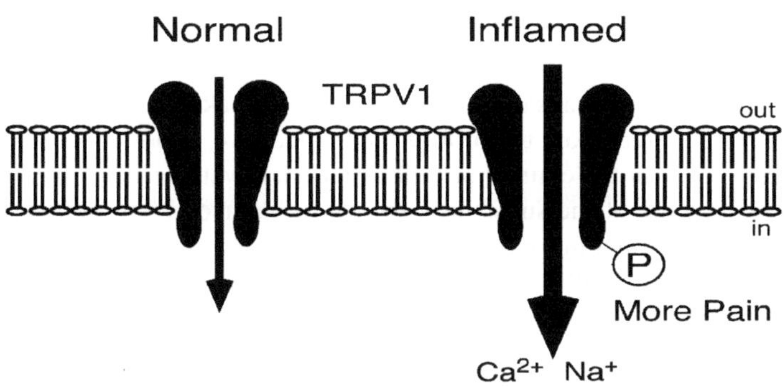

FIGURE 7.2. Proposed model for hyperalgesia induction through the phosphorylation (P) of TRPV1. Phosphorylation of TRPV1 sensitizes TRPV1 and reduces the temperature threshold for its activation, thus enhancing nociception.

may release from PIP_2-mediated inhibition of TRPV1.[35] A site within the C-terminal domain of TRPV1 that is required for PIP_2-mediated inhibition of channel gating has been identified.[36] Alternatively, lipoxygenase (LOX) products such as 12-hydroperoxyeicosatetraenooic acid (12-HPETE), anandamide and *N*-arachidonoyl dopamine (NADA) which have been reported to activate TRPV1 directly,[37–40] may also be involved in the sensitization of TRPV1 in the downstream of PLC activation, because PKC activated by DAG can activate phospholipase A_2 (PLA_2), which produces arachidonic acid (source of LOX products) from the membrane phospholipids.[41] It is not currently clear which pathways work more predominantly in the context of inflammation.

7.6. ATP and Blood Vessels

Arteries and veins receive a rich sensory innervation.[42] Sensory nerves innervating blood vessels have been implicated in the pain associated with angina, embolism, myocardial infection, and migraine, although local acidification[43,44] and released ATP have also been expected to cause nociception through the activation of ionotropic and metabotropic receptors innervating cardiac myocytes in those situations. Some reports have suggested that vasosensory nerves may be involved in reflex regulation of the cardiovascular system.[45,46] ATP, ADP, UTP, and adenosine are released from endothelial cells and platelets to act on smooth muscle $P2X_1$ receptors and on $P2X_2$ and $P2X_4$ receptors in some vessels, resulting in vasoconstriction.[47,48] They also act on smooth muscle P2Y receptors, resulting in vasodilatation.[47] Anandamide, a candidate for endogenous ligand for TRPV1, induced vasodilatation by activating TRPV1 on perivascular sensory nerves followed by release of calcitonin-gene-related peptide (CGRP).[40] Thus, TRPV1 may also be involved in the regulation of vessel tone.

7.7. Conclusion

It is now clear that the capsaicin receptor TRPV1 makes a complex with several $G_{q/11}$-coupled receptors to amplify the nociceptive information detected by TRPV1, especially in the context of inflammation. Given the fact that one of the final targets of some of the substances known to be involved in inflammation is TRPV1, compounds acting on the $G_{q/11}$-coupled receptors of the substances or TRPV1, or interfering with their interaction may prove useful in the treatment of pain and inflammation.

References

1. Fields, H.L., 1987, *Pain*, McGraw-Hill, New York.
2. Wood, J.N., and Perl, E.R., 1999, *Pain. Curr. Opin. Genet. Dev.* **9**: 328–332.
3. Cesare, P., and McNaughton, P., 1997, Peripheral pain mechanisms, *Curr. Opin. Neurobiol.* **7**: 493–499.
4. Julius, D., and Basbaum, A.I., 2001, Molecular mechanisms of nociception, *Nature* **413**: 203–210.
5. McCleskey, E.W., and Gold, M.S., 1999, Ion channels of nociception, *Annu. Rev. Physiol.* **61**: 835–856.
6. Scholz, J., and Woolf, C.J., 2002, Can we conquer pain? *Nat. Neurosci.* **5 Suppl**: 1062–1067.
7. Caterina, M.J., and Julius, D., 2001, The vanilloid receptor: a molecular gateway to the pain pathway, *Annu. Rev. Neurosci.* **24**: 487–517.
8. Caterina, M.J., Schumacher, M.A., Tominaga, M., Rosen, T.A., Levine, J.D., and Julius, D., 1997, The capsaicin receptor: a heat-activated ion channel in the pain pathway, *Nature* **389**: 816–824.
9. Tominaga, M., Caterina, M.J., Malmberg, A.B., Rosen, T.A., Gilbert, H., Skinner, K., Raumann, B.E., Basbaum, A.I., and Julius, D., 1998, The cloned capsaicin receptor integrates multiple pain-producing stimuli, *Neuron* **21**: 531–543.
10. Caterina, M.J., Leffler, A., Malmberg, A.B., Martin, W.J., Trafton, J., Petersen-Zeitz, K.R., Koltzenburg, M., Basbaum, A.I., and Julius, D., 2000, Impaired nociception and pain sensation in mice lacking the capsaicin receptor, *Science* **288**: 306–313.
11. Davis, J.B., Gray, J., Gunthorpe, M.J., Hatcher, J.P., Davey, P.T., Overend, P., Harries, M.H., Latcham, J., Clapham, C., Atkinson, K., et al., 2000, Vanilloid receptor-1 is essential for inflammatory thermal hyperalgesia, *Nature* **405**: 183–187.
12. Bodin, P., and Burnstock, G., 2001, Purinergic signalling: ATP release, *Neurochem, Res,* **26**:959–969.
13. Burnstock, G., 2001, Purine-mediated signalling in pain and visceral perception, *Trends Pharmacol. Sci.* **22**: 182–188.
14. Burnstock, G., 1999, Release of vasoactive substances from endothelial cells by shear stress and purinergic mechanosensory transduction, *J. Anat,* **194** (Pt 3): 335–342.
15. Vassort, G., 2001, Adenosine 5′-triphosphate: a P2-purinergic agonist in the myocardium, *Physiol. Rev.* **81**: 767–806.
16. Forrester, T., and Williams, C.A., 1977, Release of adenosine triphosphate from isolated adult heart cells in response to hypoxia, *J. Physiol.* **268**: 371–390.

17. Pearson, J.D., and Gordon, J.L., 1985, Nucleotide metabolism by endothelium, *Annu. Rev. Physiol.* **47**: 617–627.
18. North, R.A., and Barnard, E.A., 1997, Nucleotide receptors, *Curr. Opin. Neurobiol.* **7**: 346–357.
19. Ralevic, V., and Burnstock, G., 1998, Receptors for purines and pyrimidines, *Pharmacol. Rev.* **50**: 413–492.
20. Chen, C.C., Akopian, A.N., Sivilotti, L., Colquhoun, D., Burnstock, G., and Wood, J.N., 1995, A P2X purinoceptor expressed by a subset of sensory neurons, *Nature* **377**: 428–431.
21. Lewis, C., Neidhart, S., Holy, C., North, R.A., Buell, G., and Surprenant, A., 1995, Coexpression of P2X2 and P2X3 receptor subunits can account for ATP-gated currents in sensory neurons, *Nature* **377**: 432–435.
22. Tominaga, M., Wada, M., and Masu, M., 2001, Potentiation of capsaicin receptor activity by metabotropic ATP receptors as a possible mechanism for ATP-evoked pain and hyperalgesia, *Proc. Natl. Acad. Sci. USA* **98**: 6951–6956.
23. Numazaki, M., Tominaga, T., Toyooka, H., and Tominaga, M., 2002, Direct phosphorylation of capsaicin receptor VR1 by protein kinase $C_{epsilon}$ and identification of two target serine residues, *J. Biol. Chem.* **277**: 13375–13378.
24. Moriyama, T., Iida, T., Kobayashi, K., Higashi, T., Fukuoka, T., Tsumura, H., Leon, C., Suzuki, N., Inoue, K., Gachet, C., et al., 2003, Possible involvement of P2Y2 metabotropic receptors in ATP-induced transient receptor potential vanilloid receptor 1-mediated thermal hypersensitivity, *J. Neurosci.* **23**: 6058–6062.
25. Anderson, C.M., and Parkinson, F.E., 1997, Potential signalling roles for UTP and UDP: sources, regulation, and release of uracil nucleotides, *Trends Pharmacol, Sci,* **18**: 387–392.
26. Lazarowski, E.R., Boucher, R.C., and Harden, T.K., 2000, Constitutive release of ATP and evidence for major contribution of ecto-nucleotide pyrophosphatase and nucleoside diphosphokinase to extracellular nucleotide concentrations, *J. Biol. Chem.* **275**: 31061–31068.
27. Sawynok, J., and Sweeney, M.I., 1989, The role of purines in nociception, *Neuroscience* **32**: 557–569.
28. Molliver, D.C., Cook, S.P., Carlsten, J.A., Wright, D.E., and McCleskey, E.W., 2002, ATP and UTP excite sensory neurons and induce CREB phosphorylation through the metabotropic receptor, P2Y2, *Eur. J. Neurosci.* **16**: 1850–1860.
29. Zimmermann, K., Reeh, P.W., and Averbeck, B., 2002, ATP can enhance the proton-induced CGRP release through P2Y receptors and secondary PGE(2) release in isolated rat dura mater, *Pain* **97**: 259–265.
30. Mizumura, K., and Kumazawa, T., 1996, Modification of nociceptor responses by inflammatory mediators and second messengers implicated in their action—a study in canine testicular polymodal receptors, *Prog. Brain Res.* **113**: 115–141.
31. Sugiura, T., Tominaga, M., Katsuya, H., and Mizumura, K., 2002, Bradykinin lowers the threshold temperature for heat activation of vanilloid receptor 1, *J. Neurophysiol.* **88**: 544–548.
32. Vergnolle, N., Wallace, J.L., Bunnett, N.W., and Hollenberg, M.D., 2001, Protease-activated receptors in inflammation, neuronal signaling, and pain, *Trends Pharmacol. Sci.***22**: 146–152.
33. Amadesi, S., Nie, J., Vergnolle, N., Cottrell, G.S., Grady, E.F., Trevisani, M., Manni, C., Geppetti, P., McRoberts, J.A., Ennes, H., et al., 2004, Protease-activated receptor

2 sensitizes the capsaicin receptor transient receptor potential vanilloids receptor 1 to induce hyperalgesia, *J. Neurosci.* **24**: 4300–4312.
34. Dai, Y., Moriyama, T., Higashi, T., Togashi, K., Kobayashi, K., Yamanaka, H., Tominaga, M., and Noguchi, K., 2004, Proteinase-activated receptor 2-mediated potentiation of transient receptor potential vanilloids subfamily 1 activity reveals a mechanism for proteinase-induced inflammatory pain, *J. Neurosci.* **24**: 4293–4299.
35. Chuang, H.H., Prescott, E.D., Kong, H., Shields, S., Jordt, S.E., Basbaum, A.I., Chao, M.V., and Julius, D., 2001, Bradykinin and nerve growth factor release the capsaicin receptor from PtdIns(4,5)P2-mediated inhibition, *Nature* **411**: 957–962.
36. Prescott, E.D., and Julius, D., 2003, A modular PIP_2 binding site as a determinant of capsaicin receptor sensitivity, *Science* **300**: 1284–1288.
37. Ahern, G.P., 2003, Activation of TRPV1 by the satiety factor oleoylethanolamide, *J. Biol. Chem.* **278**: 30429–30434.
38. Huang, S.M., Bisogno, T., Trevisani, M., Al-Hayani, A., De Petrocellis, L., Fezza, F., Tognetto, M., Petros, T.J., Krey, J.F., Chu, C.J., et al., 2002, An endogenous capsaicin-like substance with high potency at recombinant and native vanilloid VR1 receptors, *Proc. Natl. Acad. Sci. USA* **99**: 8400–8405.
39. Hwang, S.W., Cho, H., Kwak, J., Lee, S.Y., Kang, C.J., Jung, J., Cho, S., Min, K.H., Suh, Y.G., Kim, D., and Oh, U., 2000, Direct activation of capsaicin receptors by products of lipoxygenases: endogenous capsaicin-like substances, *Proc. Natl. Acad. Sci. USA* **97**: 6155–6160.
40. Zygmunt, P.M., Petersson, J., Andersson, D.A., Chuang, H., Sorgard, M., Di Marzo, V., Julius, D., and Hogestatt, E.D., 1999, Vanilloid receptors on sensory nerves mediate the vasodilator action of anandamide, *Nature* **400**: 452–457.
41. Shin, J., Cho, H., Hwang, S.W., Jung, J., Shin, C.Y., Lee, S.Y., Kim, S.H., Lee, M.G., Choi, Y.H., Kim, J., et al., 2002, Bradykinin-12-lipoxygenase-VR1 signaling pathway for inflammatory hyperalgesia, *Proc. Natl. Acad. Sci. USA* **99**: 10150–10155.
42. Coleridge, H.M., and Coleridge, J.C., 1980, Cardiovascular afferents involved in regulation of peripheral vessels, *Annu. Rev. Physiol.* **42**: 413–427.
43. Malliani, A., Lombardi, F., and Pagani, M., 1981, Functions of afferents in cardiovascular sympathetic nerves, *J. Auton. Nerv. Syst.* **3**: 231–236.
44. Moskowitz, M.A., 1990, Basic mechanisms in vascular headache, *Neurol. Clin.* **8**: 801–815.
45. Benson, C.J., Eckert, S.P., and McCleskey, E.W., 1999, Acid-evoked currents in cardiac sensory neurons: a possible mediator of myocardial ischemic sensation, *Circ, Res,* **24**: 921–928.
46. Vaishnava, P., and Wang, D.H., 2003, Capsaicin sensitive-sensory nerves and blood pressure regulation, *Curr. Med. Chem. Cardiovasc. Hematol. Agents* **1**: 177–188.
47. Burnstock, G., 2002, Purinergic signalling and vascular cell proliferation and death, *Arterioscler. Thromb. Vasc. Bio* **22**: 364–373.
48. Kunapuli, S.P., and Daniel, J.L., 1998, P2 receptor subtypes in the cardiovascular system, *Biochem. J.* **336** (Pt 3): 513–523.

8
TRPV4 and Hypotonic Stress

David M. Cohen*

Abstract: TRPV4 was identified as the mammalian homologue of the *C. elegans* osmosensory channel protein, OSM-9. This nonselective cation channel is activated *in vitro* by even modest degrees of hypotonic cell swelling. Its expression in the mammalian central nervous system and kidney suggests a role in systemic osmoregulation—a view borne out by detailed balance studies in TRPV4-null mice. Two distinct mechanisms have been described through which TRPV4 may be activated by hypotonicity: one involves the SRC family of nonreceptor protein tyrosine kinases, while the other is mediated via arachidonic acid metabolites.

8.1. The TRP Channel Superfamily

Channels of the TRP family are responsive to a wide range of both intracellular and extracellular stimuli. These nonselective cation channels are named for their founding member, *trp* ("transient receptor potential"),[1,2] which when mutated in *Drosophila* gives rise to aberrant photoreceptor activity.[3] TRPC (<u>c</u>lassical) channels are activated following ligand engagement of either receptor tyrosine kinases or G-protein-coupled receptors.

TRPC channel subunits, perhaps in the form of heterotetramers, likely account for the store-operated calcium channel activity observed in some model systems. TRPM (<u>m</u>elastatin) channels uniquely possess long intracellular carboxy termini which, in several members of this family, encode functional kinase domains. TRPV (<u>v</u>anilloid) channels are responsive to environmental stimuli. Of the six known family members, TRPV1, −2, −3, and −4 all respond to elevated temperature *in vitro*, albeit with differing thresholds. TRPV2 is activated by the highest temperature,[4] followed by TRPV1, TRPV3, and TRPV4. Of note, TRPV1 was first cloned as the receptor for the irritant and active ingredient in "hot" peppers, capsaicin.[5]

*Division of Nephrology and Hypertension, Oregon Health and Science University and the Portland Veterans Affairs Medical Center, Portland, OR 97239, cohend@ohsu.edu

8.2. OSM-9 and the Cloning of TRPV4

The still-unfolding story of TRPV4 and osmoregulation began in C. elegans. Because the nematodes avoid steep osmotic gradients to prevent acute changes in cell volume, this simple organism provided a tractable genetic model system for studying osmotic sensing. Colbert et al. noted that worms with mutations in a particular gene, while fully motile, no longer avoided potentially harmful osmotic gradients; this gene was dubbed *osm-9*.[6,7] Early work had focused on the potential chemosensory role of the OSM-9 protein in odorant signal transduction.

The principal chemosensory organs in C. elegans, the amphids, are comprised of twelve pairs of neurons. Ten of these neuronal pairs express the OSM-9 protein. This protein bore a striking similarity to members of the transient receptor potential family of channels in Drosophila and mammals; in particular, it was reminiscent of what later became known as the vanilloid-responsive TRPV channels. Four other TRPV-like channels were later identified in *C. elegans*. These are distributed among the amphid, and other, sensory neurons, often in conjunction with OSM-9.[8] In these locations, the worm TRPV analogs contribute to odorant detection and mechanosensation. Interestingly, targeting expression of the vanilloid-responsive mammalian TRPV1 to the appropriate *C. elegans* chemosensory nuclei renders the resultant mutant newly sensitive to the vanilloid, capsaicin.[8] Such module-swapping experiments serve to underscore the phylogenetic conservation of TRP-dependent sensory pathways.

Mammalian TRPV4 was cloned as the ortholog of *C. elegans* OSM-9.[9,10] When heterologously expressed, TRPV4 was activated by hypotonicity; a decrement in tonicity of as little as one percent was sufficient for this activation.[9,10]

8.3. Activation of TRPV4 by Thermal Stress and Lipids

Mammalian TRPV4 was described in the context of cell response to hypotonicity at the cellular and organismal levels and thus remains the most widely studied stimulus; other activators, however, have emerged. As described above, TRPV4 is activated by thermal stress. This phenomenon has been observed under native conditions in keratinocytes (skin cells)[11] and aortic endothelial cells,[15] and following heterologous expression in HEK cells and *Xenopus* oocytes.[13–15] Functional redundancy in thermal sensation among TRPV4 and related TRPV channels has been incompletely explored. For example, mice lacking TRPV4 exhibit the same latency to gradual temperature elevation on a heated plate as control mice;[16] see below.

TRPV4 is also activated via a variety of lipids. Although unresponsive to the vanilloid compounds and TRPV1 agonists, capsaicin and resiniferatoxin,[9,10] TRPV4 was activated by the phorbol ester and activator of protein kinases C (PKCs), phorbol 12-myristate 13-acetate (PMA).[12] Control experiments performed with the non-PKC-activating PMA analog, 4α-phorbol 12,13-didecanoate (4α-PDD) yielded a surprising result: this "inactive" control compound was fifty times more potent than PMA.[12] In addition, 4α-PDD failed to activate other TRPV family members, indicating a degree of specificity for TRPV4.[17]

Following a search for the true "physiological" lipid agonist of TRPV4, this group made the striking observation that the endogenous cannabinoid anandamide activated the channel.[18] Although cannabinoids were known to be potent agonists of TRPV1 (reviewed in Ref. 17), their role in TRPV4 activation had not been explored. Classically, cannabinoids exert their myriad physiological effects via interaction with specific G-protein-coupled receptors,[19] but this effect of anandamide did not require the participation of cannabinoid receptors. Through the use of pharmacological inhibitors and agonists, Watanabe et al. suggested that anandamide is not a direct activator of TRPV4. Rather, anandamide is likely metabolized to arachidonic acid and then, via the action of cytochrome P450 epoxygenase, is metabolized further to one of several epoxyeicosatrienoic acid (EET) compounds. These lipids, particularly 5′,6′-epoxyeicosatrienoic acid, then activate TRPV4. Whether activation of TRPV4 by these lipid compounds is direct or indirect remains unresolved. Exogenously applied EETs activate TRPV4.[18] But there are myriad catabolic pathways for the EETs;[20] much as the TRPV4 activating ability of arachidonic acid was traced to its EET metabolites, so might these putative effects of EET be in fact attributable to one of its many metabolites. In addition, a variety of lipids including arachidonic acid and anandamide influence ion channel behavior indirectly by altering the cell membrane lipid microenvironment where the channel resides (e.g., Refs. 21, 22).

8.4. Molecular Mechanism of TRPV4 Activation by Hypotonicity

Hypotonicity is the best-studied context for activation of TRPV4. Conflicting data have emerged, however, with respect to the molecular mechanism through which this stimulus influences TRPV4 function. Xu et al. noted that both heterologously expressed and natively expressed TRPV4 undergo tyrosine phosphorylation shortly after cells are exposed to hypotonic stress. In the murine distal convoluted tubule cell line, mDCT,[23] derived from a kidney tubule segment where TRPV4 is known to be expressed, this effect was evident within one minute.[24] The effect appeared to be specific to hypotonic stress; it was not reproduced by phorbol ester or by peptide growth factor application. The effect was largely sensitive to PP1, an inhibitor of Src-family cytoplasmic tyrosine kinases. In addition, a range of Src family kinases associated with natively expressed and heterologously expressed TRPV4, as determined by co-immunoprecipitation and co-localization experiments. Overexpression of one of these kinases, Lyn, enhanced both basal and hypotonicity-inducible tyrosine phosphorylation of TRPV4, whereas overexpression of dominant-negative acting Lyn partially blocked the response. Hypotonicity activated Lyn , and recombinant Lyn protein phosphorylated recombinant TRPV4 *in vitro*. Systematic mutation of tyrosine residues in TRPV4 indicated that Tyr-253 was the site of regulated phosphorylation. Mutation of this site to any of a number of different residues blocked hypotonicity-dependent calcium transients, even though the mutant channel was abundantly expressed[24] and was appropriately targeted to the plasma membrane as determined via cell surface biotinylation studies.

Other data support a role for tyrosine phosphorylation in signaling by anisotonicity. Tyrosine kinase activity is upregulated by hypotonic stress,[25–27] and pharmacological manipulation of global tyrosine kinase or phosphatase activity influences regulatory volume decrease-associated cell efflux of ions and amino acid osmolytes.[27–29] In addition, a number of SRC family kinases are activated by cell swelling and have been implicated in the physiological response to hypotonic stress.[30] Furthermore, the Src-family kinase Fyn particpates in regulation of the tonicity-responsive transcription factor, TonEBP.[31] Moreover, it was subsequently shown that, in addition to TRPV4, other TRP channels including members of the TRPC, TRPM, and TRPV families are similarly regulated by SRC family kinases.[32–35]

Multiple pathways potentially mediate activation of SRC family kinases by hypotonic stress and are the subject of a recent review;[30] see Fig 8.1. Integrins, mediating cell-substrate and cell-cell interactions, are well positioned to sense the membrane deformation accompanying acute changes in cell volume.[36] Integrins likely associate with membrane-spanning ligands for receptor tyrosine kinases such as the epidermal growth factor (EGF) receptor. One such example is the EGF receptor ligand, heparin-binding epidermal growth factor (HB-EGF). Activation of the EGF receptor via stress-dependent cleavage of the active "ectodomain" of HB-EGF (and related ligands) triggers multiple downstream signaling events that converge on SRC and related kinases (Fig 8.1).

In contrast, Nilius and colleagues argue that SRC family kinase activity and tyrosine phosphorylation are not required for activation of TRPV4 by hypotonicity. They transiently expressed Tyr-253-mutant TRPV4 in HEK293 cells and were still able to demonstrate hypotonicity-induced calcium transients.[37] These authors used a bicistronic expression vector (resulting in co-expression of GFP), a transient transfection approach, and a single-cell fura-2-based calcium assay in adherent cells. Xu et al, in contrast, relied on stable transfection of a monocistronic expression vector, and performed their fura-2-based calcium assay using cells in suspension.[24] The Nilius group also used pharmacological inhibitors of phospholipase A_2 to suggest that this enzyme was pivotal to the hypotonic activation of TRPV4. PLA_2 inhibition with the compound pBPB, for example, prevented TRPV4 activation in response to hypotonic stress but not in response to the lipid agonist, 4α-PDD.[37] These data were consistent with earlier findings that hypotonicity activated phospholipase A_2-dependent arachidonic acid release.[38–40]

Just as anandamide could be metabolized through arachidonic acid to the active epoxyeicosatrienoic acids,[18] so could membrane phospholipids be metabolized to arachidonic acid via PLA_2 and thereby feed into the same downstream cascade (Fig. 8.1). Unexpectedly, cytochrome P450 epoxygenase—which functions downstream of phospholipase A_2 and is required for TRPV4 activation by the cannabinoid anandamide (*v.s.*)—is not a universal participant among TRPV4-activating

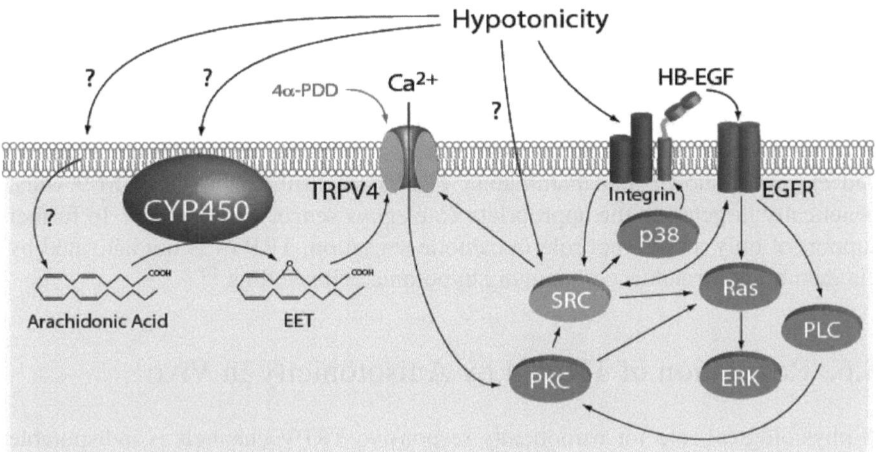

FIGURE 8.1. Mechanism of activation of TRPV4 in response to hypotonic cell swelling. Summary of signaling events culminating in activation of TRPV4 by cell swelling. Hypotonicity directly or indirectly activates SRC (here representing either SRC or another member of the SRC family of nonreceptor protein tyrosine kinases), perhaps via an integrin-dependent pathway. Some data support a role for activation of the epidermal growth factor receptor (EGFR) via metalloproteinase-dependent ectodomain cleavage of EGFR ligands such as HB-EGF. Activation of EGFR may directly or indirectly lead to activation of SRC family kinases (SFKs). Active SFKs then phosphorylate Tyr-253 on TRPV4, permitting calcium entry. Hypotonicity also activates phospholipase A_2 which cleaves membrane lipids to liberate arachidonic acid. Arachidonic acid is then converted to one of a series of epoxyeicosatrienoic acid (EET) compounds via membrane-associated cytochrome P450 epoxygenase activity. EETs, as well as the synthetic lipid agonist, 4α-PDD (or their further metabolites) activate TRPV4. Alternatively, these compounds may alter the lipid milieu of TRPV4 (indicated by arrows entering cell membrane), indirectly influencing gating of the channel. "SRC" denotes either SRC itself or a related SRC family kinase. Arrows signify either direct or indirect activation events. In general, although not conveyed by this diagram in the interest of clarity, activated SRC physically associates with activated integrins and EGFR in the cell membrane. (See Color Plate 11 in Color Section)

pathways. Inhibition of this enzyme with miconazole or with 17-ODYA effectively blocked calcium entry in response to hypotonic stress but not in response to heat stress.[37] These authors also went on to implicate involvement of another tyrosine residue, Tyr-555, in activation of TRPV4 by heat and 4α-PDD but, importantly, not by cell swelling.[37]

8.5. Is TRPV4 an Actual Osmotic Sensor?

Data from Ref. 7 and others indicate that the prototypical TRPV4 channel, OSM-9, is potentially responsive not only to osmotic stress, but also to mechanosensory and chemosensory stimuli. For example, OSM-9-deficient nematodes also lost the ability to sense the olfactory agonists diacetyl and pyrazine but not other odorants. In the worm olfactory response, OSM-9 likely functions downstream of a G-protein-coupled receptor. This hierarchical arrangement suggests that rather than constituting an osmotic sensor per se, TRPV4 might represent an osmotic effector. Facilitating extrapolation of these data across divergent species, it was recently and elegantly shown that mammalian TRPV4 can substitute for OSM-9 when genetically targeted to the appropriate *C. elegans* sensory apparatus.[41] In further support of only an indirect role in osmotic sensation, TRPV4 is not activated by the membrane stretch accompanying hypotonic cell swelling.[10]

8.6. Regulation of TRPV4 by Anisotonicity In Vivo

A physiological role for osmotically responsive TRPV channels is indisputable in *C. elegans*; among higher eukaryotes, data are emerging more slowly. Initial speculation was fueled by the attractive localization of TRPV4 expression to key osmosensory and mechanosensory tissues,[9,14] although some of these observations may have reflected selection bias on the part of investigators. Nonetheless, TRPV4 is expressed in the mechanosensitive neurons of the mammalian inner ear hair cells, and in the osmo-sensing blood-brain barrier-deficient circumventricular nuclei.[9] Their expression in the latter context suggested a role in systemic, rather than cellular, osmoregulation.

To test this in a rigorous fashion, Mizuno and colleagues generated a TRPV4-null mouse model.[42] Unexpectedly, there was no gross perturbation in systemic salt or water balance. There was a trend toward higher plasma sodium concentration (which is generally indicative of greater water loss) in the TRPV4-null animals, although statistical significance was not achieved with a limited number of mice per group. Similarly inconclusive findings were evident following provocative testing with either water or Na^+ loading. Interestingly, although the TRV4-null and wild-type mice exhibited comparable basal serum levels of the water-retentive hormone arginine vasopressin, the response to simultaneous water restriction and osmotic loading (in the form of intraperitoneal propylene glycol) was much more robust in the TRPV4 –/– mice. This enhanced AVP response to a hyperosmolar state was recapitulated in vitro using brain slices.[42]

Liedtke and colleagues independently generated a TRPV4-null mouse model,[43] and when mice were housed in groups and had unlimited access to food and water, there was no difference in plasma osmolality between wild-type and TRPV4-null groups. However, reasoning that mice, like humans, respond to social cues to water consumption, they next housed their animals singly and repeated the experiment. (Also, rather than measuring plasma [Na^+], this group measured plasma osmolality, an independent index of plasma water content that generally but not invariably tracks with plasma [Na^+]). In their hands, the TRPV4 –/– mice exhibited significantly higher plasma osmolality (300 rather than 295 mOsmol/kgH_2O). In corresponding fashion, the knock-out mice also drank $\sim$ 30% less water in the first 24 hours of isolation. Oddly, whereas the Mizuno group showed increased

AVP response to hyperosmotic challenge,[42] the Liedtke group demonstrated a *less* robust plasma AVP response in this setting, albeit with a different osmotic stimulus (intraperitoneal administration of 0.5 M NaCl).[43]

Functional discrepancies between these two knockout models may be explained in part by differences in the way they were generated. The model of Liedtke et al. incorporated a resistance cassette in an exon near the carboxy-terminus of the TRPV4 gene.[43] The mice of Mizuno et al., in contrast, were created by insertion of a neomycin-resistant gene in an amino-terminal exon. These latter mice still expressed a truncated TRPV4 mRNA, which was initially felt to be devoid of functional activity.[44,45] Were this splice variant to be natively expressed in wild-type mice, as some suggest to be the case,[16] it might serve some as-of-yet unknown function and thereby confound data interpretation.

TRPV4 is highly expressed in mammalian kidney;[9,10] initially reported expression in the distal convoluted tubule. In a typical person, the kidneys convert ~150 liters of daily glomerular filtrate (or "potential urine") into 0.5–1.5 liters of urine; in so doing, they also reabsorb nearly all of the filtered water and inorganic ions, as well as glucose, amino acids, and a wide variety of essential plasma constituents.

The 100-fold reduction in urine volume accomplished along the kidney tubule is clearly a pivotal function; in its absence, lethal volume depletion and hypotension would ensue in minutes. This robust water conservation is achieved via a combination of apical water permeability of specific kidney tubule segments, and a hyperosmotic renal medullary interstitium. Together, these factors create both a route and a driving force for water reabsorption into the systemic circulation.

Although no data specifically confirm a role for TRPV4 in this process, clues have emerged from a detailed investigation of the pattern of TRPV4 expression in the kidney. Glomerular filtrate traverses the hyperosmotic renal medulla via the tight hairpin-like loop of Henle. The cells lining the "descending" arm of the loop are freely water permeable; the filtrate becomes progressively more concentrated as water traverses the tubule cells and enters the hyperosmotic interstitium. An abrupt transition occurs at the genu of the loop, where the "descending" limb meets the "ascending" limb: here apical water permeability abruptly ceases. This architecture "traps" the newly concentrated filtrate in the lumen, preventing its passive dilution by water entry as it exits the hypertonic inner medulla of the kidney.

TRPV4 expression is completely absent along the early parts of the kidney tubule where passive water reabsorption takes place; however, at the precise transition between tubule water permeability and impermeability (at the hairpin turn at the base of the loop of Henle), TRPV4 expression abruptly emerges (Fig. 8.2). And TRPV4 expression continues for the length of the kidney tubule. The solitary exception is the highly specialized tubule segment called the macula densa, the only other tubule segment exhibiting constitutive apical water permeability. Here, as in the earlier water-permeable tubule segments, TRPV4 expression is absent.[46] Therefore, TRPV4 expression in the kidney is completely restricted to those tubule segments that lack constitutive apical water permeability, and where a transcellular osmotic gradient may be expected to develop.

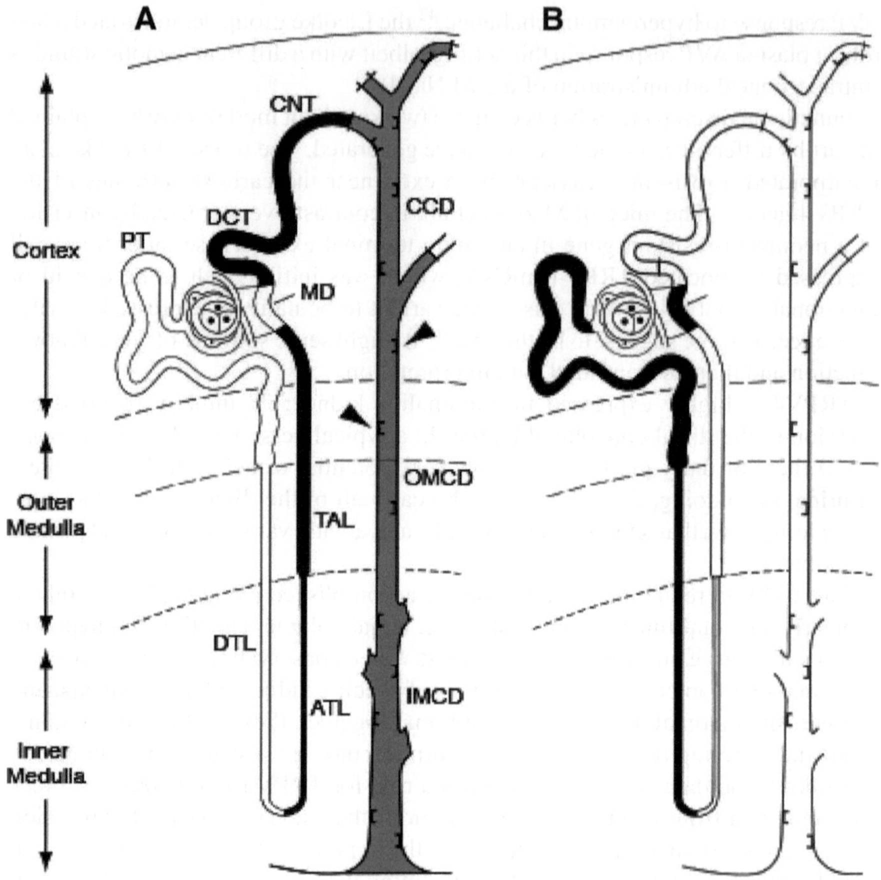

FIGURE 8.2. TRPV4 expression in the mammalian kidney. (A) TRPV4 expression along the mouse and rat nephron. A single nephron is shown. Black shading denotes strong TRPV4 expression; gray shading indicates moderate TRPV4 expression. Throughout the collecting duct, expression was stronger in the intercalated cells (e.g., arrowheads). The kidney is divided into cortex and medulla; the latter consists of outer and inner medulla. Outer medulla is further divided into outer stripe and inner stripe (not labeled). PT, proximal tubule; DTL, descending thin limb of the loop of Henle; ATL, ascending thin limb of the loop of Henle; MD, macula densa; DCT, distal convoluted tubule; CNT, connecting tubule; CCD, cortical collecting duct; OMCD, outer medullary collecting duct; IMCD, inner medullary collecting duct. (B) Apical water permeability along the nephron. Black shading denotes tubule segments with constitutive apical water permeability. Note that this distribution is the inverse of that of TRPV4 expression in panel A. (Modified from Ref. 46.)

In summary, TRPV4 is activated by arachidonic acid metabolites, and by a variety of physical stimuli including thermal and osmotic stresses. It is likely that tonicity-dependent activation of TRPV4 requires SRC family kinase-dependent tyrosine phosphorylation in at least some experimental contexts, and data support a role for phospholipase A_2 and cytochrome P450 epoxygenase activity and involvement of epoxyeicosatrienoic acids. Importantly, what relationship, if any, exists between these two pathways, remains incompletely explored. Similarly unclear is the precise role played by both centrally expressed (i.e., hypothalamic) and peripherally expressed (i.e., renal) TRPV4 in influencing systemic water balance.

References

1. Montell, C., Jones, K., Hafen, E., and Rubin, G., 1985, Rescue of the *Drosophila* phototransduction mutation trp by germline transformation, *Science* **230**: 1040–1043.
2. Montell, C., and Rubin, G.M., 1989, Molecular characterization of the *Drosophila* trp locus: a putative integral membrane protein required for phototransduction, *Neuron* **2**: 1313–1323.
3. Minke. B., 1977, *Drosophila* mutant with a transducer defect, *Biophys. Struct. Mech.* **3**: 59–64.
4. Clapham, D.E., 2003, TRP channels as cellular sensors, *Nature* **426**: 517–524.
5. Caterina, M.J., Schumacher, M.A., Tominaga M., Rosen T.A., Levine J.D., and Julius D., 1997, The capsaicin receptor: a heat-activated ion channel in the pain pathway, *Nature* **389**: 816–824.
6. Colbert, H.A., and Bargmann, C.I., 1995, Odorant-specific adaptation pathways generate olfactory *plasticity in C. elegans, Neuron* **14**: 803–812.
7. Colbert, H.A., Smith, T.L., and Bargmann, C.I., 1997, OSM-9, a novel protein with structural similarity to channels, is required for olfaction, mechanosensation, and olfactory adaptation in *Caenorhabditis elegans, J. Neurosc,i* **17**: 8259–8269.
8. Tobin, D., Madsen, D., Kahn-Kirby, A., Peckol, E., Moulder, G., Barstead, R., Maricq A., and Bargmann, C., 2002, Combinatorial expression of TRPV channel proteins defines their sensory functions and subcellular localization in *C. elegans* neurons, *Neuron* **35**: 307–318.
9. Liedtke, W., Choe, Y., Marti-Renom, M.A., Bell, A.M., Deni,s C.S., Sali, A., Hudspeth, A.J., Friedman, J.M., and Heller, S., 2000, Vanilloid receptor-related osmotically activated channel (VR-OAC), a candidate vertebrate osmoreceptor, *Cell* **103**: 525–535.
10. Strotmann, R., Harteneck, C., Nunnenmacher, K., Schultz, G., and Plant, T.D., 2000, OTRPC4, a nonselective cation channel that confers sensitivity to extracellular osmolarity, *Nat. Cell Biol.* **2**: 695–702.
11. Chung, M.K, Lee, H., and Caterina, M.J., 2003, Warm temperatures activate TRPV4 in mouse 308 keratinocytes, *J. Biol. Chem.* **278**: 32037–32046.
12. Watanabe, H., Davis, J.B., Smart, D., Jerman, J.C., Smith, G.D., Hayes, P., Vriens, J., Cairns, W., Wissenbach, U., Prenen, J., Flockerzi, V., Droogmans, G., Benham, C.D., and Nilius, B., 2002a, Activation of TRPV4 channels (hVRL-2/mTRP12) by phorbol derivatives, *J. Biol. Chem.* **277**: 13569–13577.
13. Gao X., Wu L., and O'Neil, R.G., 2003, Temperature-modulated diversity of TRPV4 channel gating: Activation by physical stresses and phorbol ester derivatives through protein kinase C-dependent and -independent pathways, *J. Biol. Chem.* **278**: 27129–27137.
14. Guler, A.D., Lee, H., Iida, T., Shimizu, I., Tominaga, M., and Caterina, M., 2002, Heat-evoked activation of the ion channel, TRPV4, *J. Neurosci.* **22**: 6408–6414.

15. Watanabe, H., Vriens, J., Suh, S.H., Benham, C.D., Droogmans, G., and Nilius, B., 2002b, Heat-evoked activation of TRPV4 channels in a HEK293 cell expression system and in native mouse aorta endothelial cells, *J. Bio.l Chem.* **277**: 47044–47051.
16. Todaka, H., Taniguchi, J., Satoh, J., Mizuno, A., and Suzuki, M., 2004, Warm temperature-sensitive transient receptor potential vanilloid 4 (TRPV4) plays an essential role in thermal hyperalgesia, *J. Biol. Chem.* **279**: 35133–35138.
17. Nilius B., Vriens J., Prenen J., Droogmans G., and Voets T., 2004, TRPV4 calcium entry channel: a paradigm for gating diversity, *Am. J. Physiol. Cell Physiol.* **286**: C195–205.
18. Watanabe, H., Vriens, J., Prenen, J., Droogmans, G., Voets, T., and Nilius, B., 2003, Anandamide and arachidonic acid use epoxyeicosatrienoic acids to activate TRPV4 channels, *Nature* **424**: 434–438.
19. Piomelli, D., 2003, The molecular logic of endocannabinoid signalling, *Nat. Rev. Neurosci.* **4**: 873–884.
20. Zeldin, D.C., 2001, Epoxygenase pathways of arachidonic acid metabolism, *J. Biol. Chem.* **276**: 36059–36062.
21. Hilgemann, D.W., 2004, Biochemistry: oily barbarians breach ion channel gates, *Science* **304**: 223–224.
22. Oliver, D., Lien, C.C., Soom, M., Baukrowitz, T., Jonas, P., and Fakler, B., 2004, Functional conversion between A-type and delayed rectifier K+ channels by membrane lipids, *Science* **304**: 265–270.
23. Gesek, F.A., and Friedman, P.A., 1992, Mechanism of calcium transport stimulated by chlorothiazide in mouse distal convoluted tubule cells, *J. Clin. Invest.* **90**: 429–438.
24. Xu, H., Zhao, H., Tian, W., Yoshida, K., Roullet, J.B., and Cohen, D.M., 2003, Regulation of a transient receptor potential (TRP) channel by tyrosine phosphorylation. SRC family kinase-dependent tyrosine phosphorylation of TRPV4 on TYR-253 mediates its response to hypotonic stress, *J. Biol. Chem.* **278**: 11520–11527.
25. Edashige, K., Watanabe, Y., Sato E.F., Takehara Y., and Utsumi, K., 1993, Reversible priming and protein-tyrosyl phosphorylation in human peripheral neutrophils under hypotonic conditions, *Arch. Biochem. Biophys.* **302**: 343–347.
26. Sadoshima, J., Qiu, Z.H., Morgan, J.P., and Izumo, S., 1996, Tyrosine kinase activation is an immediate and essential step in hypotonic cell swelling-induced ERK activation and c-fos gene expression in cardiac myocytes, *Embo. J.* **15**: 5535–5546.
27. Tilly, B.C., van den Berghe, N., Tertoolen, L.G., Edixhoven, M.J., and de Jonge, H.R., 1993, Protein tyrosine phosphorylation is involved in osmoregulation of ionic conductances, *J. Biol. Chem.* **268**: 19919–19922.
28. Good, D.W., 1995, Hyperosmolality inhibits bicarbonate absorption in rat medullary thick ascending limb via a protein-tyrosine kinase-dependent pathway, *J. Biol. Chem,* **270**: 9883–9889.
29. Sachs, J.R., and Martin, D.W., 1993, The role of ATP in swelling-stimulated K-Cl cotransport in human red cell ghosts: phosphorylation-dephosphorylation events are not in the signal transduction pathway, *J. Gen. Physio.l* **102**: 551–573.
30. Cohen, D., 2006, SRC family kinases in cell volume regulation, *Am. J. Physiol,* **288**: 6483–6493.
31. Ko, B.C, Lam, A.K., Kapus, A., Fan, L., Chung, S.K., and Chung, S.S., 2002, Fyn and p38 signaling are both required for maximal hypertonic activation of the osmotic response element-binding protein/tonicity-responsive enhancer-binding protein (OREBP/TonEBP), *J. Biol. Chem.* **277**: 46085–46092.

32. Hisatsune, C., Kuroda, Y., Nakamura, K., Inoue, T., Nakamura, T., Michikaw,a T., Mizutani, A., and Mikoshiba K., 2004, Regulation of TRPC6 channel activity by tyrosine phosphorylation, *J. Biol. Chem.* **279**: 18887–18894.
33. Jiang, X., Newell, E.W., and Schlichter, L.C., 2003, Regulation of a TRPM7-like current in rat brain microglia, *J. Biol. Chem.* **278**: 42867–42876.
34. Jin, X., Morsy, N., Winston, J., Pasricha, P.J., Garrett, K., and Akbarali, H.I., 2004, Modulation of TRPV1 by nonreceptor tyrosine kinase, c-Src kinase, *Am. J. Physiol. Cell Physiol.* **287**: C558–563.
35. Vazquez, G., Wedel, B.J., Kawasaki, B.T., Bird, G.S., and Putney, J.W., Jr., 2004, Obligatory role of Src kinase in the signaling mechanism for TRPC3 cation channels, *J. Biol. Chem.* 279: 40521–40528.
36. Alenghat, F.J., and Ingber D.E., 2002, Mechanotransduction: all signals point to cytoskeleton, matrix, and integrins, *Sci. STKE* **2002**: E6.
37. Vriens J., Watanabe H., Janssens A., Droogmans G., Voets T., and Nilius B., 2004, Cell swelling, heat, and chemical agonists use distinct pathways for the activation of the cation channel TRPV4, *Proc. Natl. Acad. Sci. USA* **101**: 396–401.
38. Basavappa, S., Pedersen, S.F., Jorgensen, N.K., Ellory, J.C., and Hoffmann, E.K., 1998, Swelling-induced arachidonic acid release via the 85-kDa cPLA2 in human neuroblastoma cells, *J. Neurophysio.l* **79**: 1441–1449.
39. Pedersen, S., Lambert, I.H., Thoroed, S.M., and Hoffmann, E.K., 2000, Hypotonic cell swelling induces translocation of the alpha isoform of cytosolic phospholipase A2 but not the gamma isoform in Ehrlich ascites tumor cells, *Eur. J. Biochem.* **267**: 5531–5539.
40. Thoroed, S.M., Lauritzen, L., Lambert, I.H., Hansen, H.S., and Hoffmann, E.K., 1997, Cell swelling activates phospholipase A2 in Ehrlich ascites tumor cells, *J. Membr. Biol.* **160**: 47–58.
41. Liedtke, W., Tobin, D.M., Bargmann, C.I., and Friedman, J.M., 2003, Mammalian TRPV4 (VR-OAC) directs behavioral responses to osmotic and mechanical stimuli in *Caenorhabditis elegans*, *Proc. Natl. Acad. Sci. USA* **27**: 27.
42. Mizuno, A., Matsumoto, N., Imai, M., and Suzuki, M., 2003, Impaired osmotic sensation in mice lacking TRPV4, *Am. J. Physiol. Cell Physiol.* **285**: C96–C101.
43. Liedtke, W., and Friedman, J.M., 2003, Abnormal osmotic regulation in trpv4−/− mice, *Proc. Natl. Acad. Sci. USA* **100**: 13698–13703.
44. Chung, M.K, Lee, H., Mizuno, A., Suzuki, M., and Caterina, M.J., 2004, TRPV3 and TRPV4 mediate warmth-evoked currents in primary mouse keratinocytes, *J. Biol. Chem.* **279**: 21569–21575.
45. Suzuki, M., Hirao, A., and Mizuno, A., 2003, Microfilament-associated protein 7 increases the membrane expression of transient receptor potential vanilloid 4 (TRPV4), *J. Biol. Chem.* **278**: 51448–51453.
46. Tian, W., Salanova, M., Xu, H., Lindsley, J.N., O,yama T.T., Anderson, S., Bachmann, S., and Cohen, D.M., 2004, Renal expression of osmotically responsive cation channel TRPV4 is restricted to water-impermeant nephron segments, *Am. J. Physiol. Renal Physiol.* **287**: F17–24.

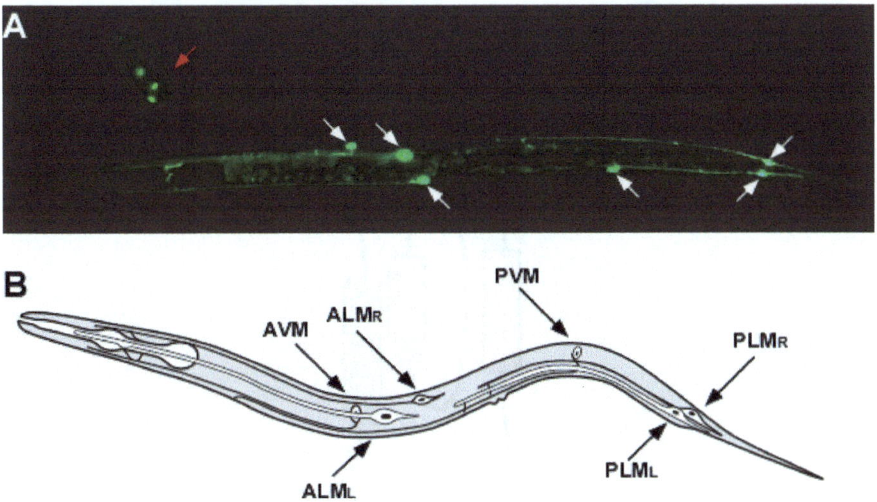

COLOR PLATE 1.

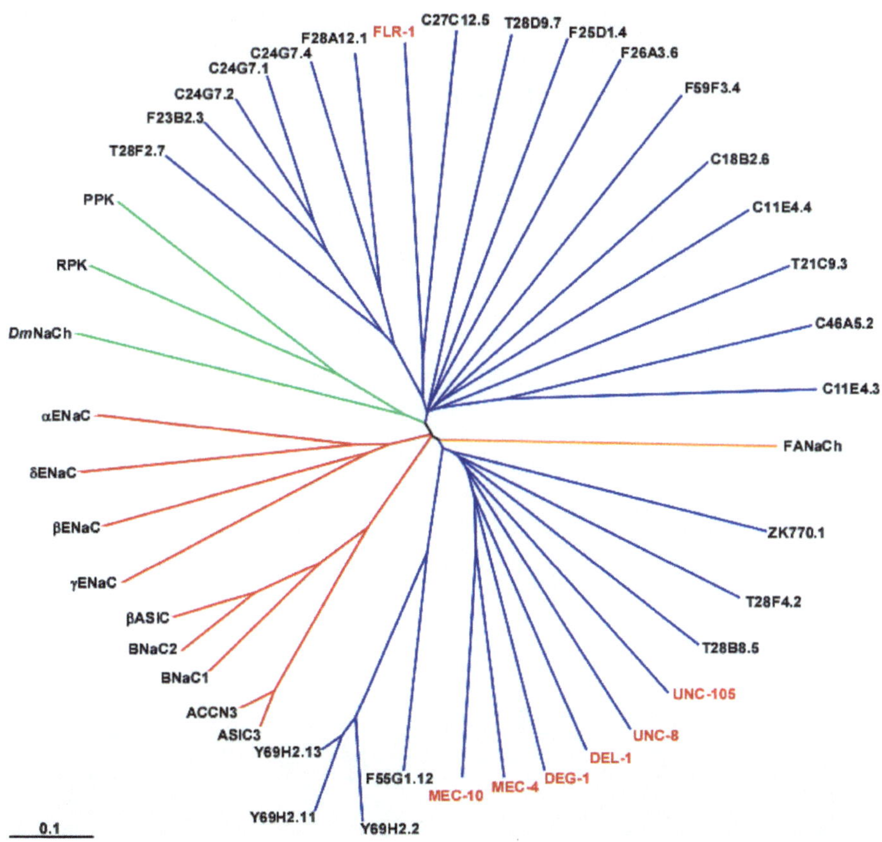

COLOR PLATE 2.

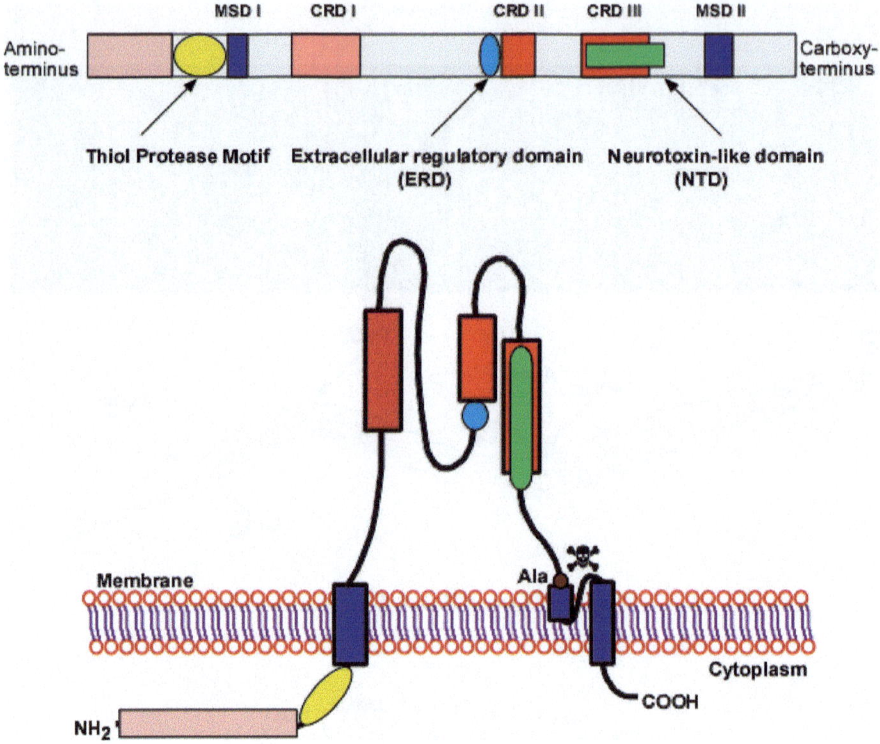

Color Plate 3.

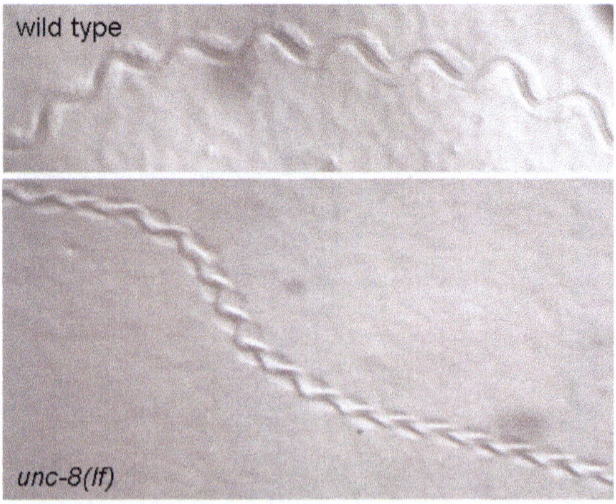

Color Plate 4.

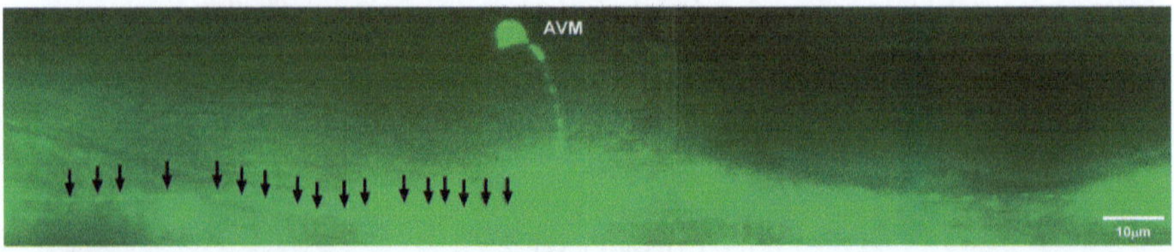

Color Plate 5.

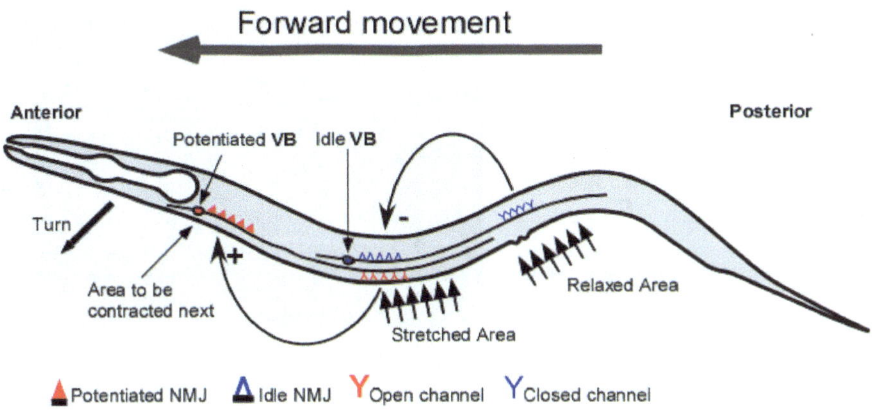

COLOR PLATE 6.

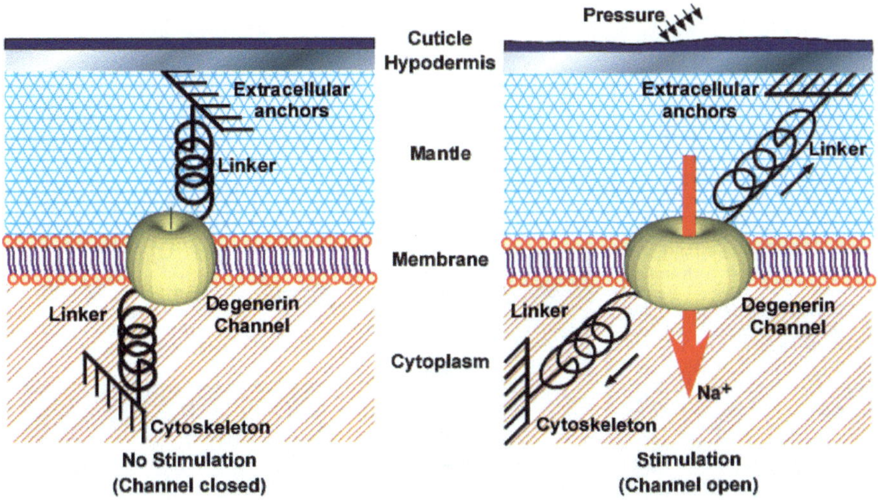

COLOR PLATE 7.

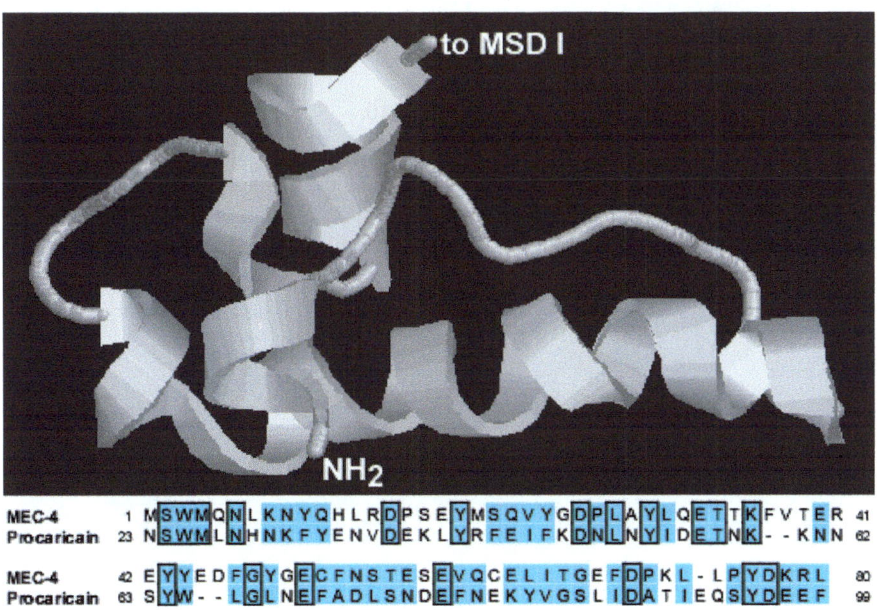

COLOR PLATE 8.

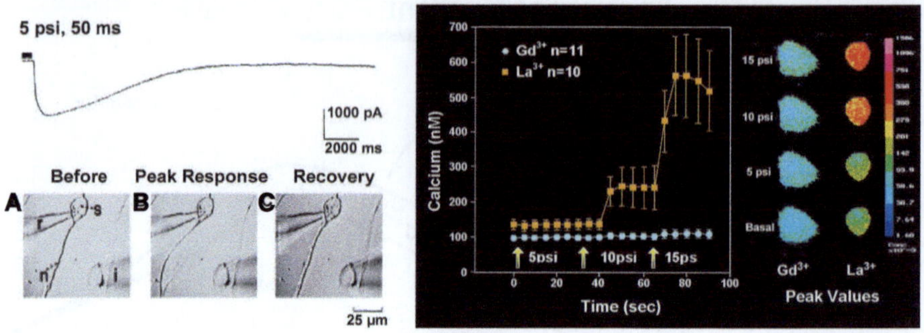

Color Plate 9.

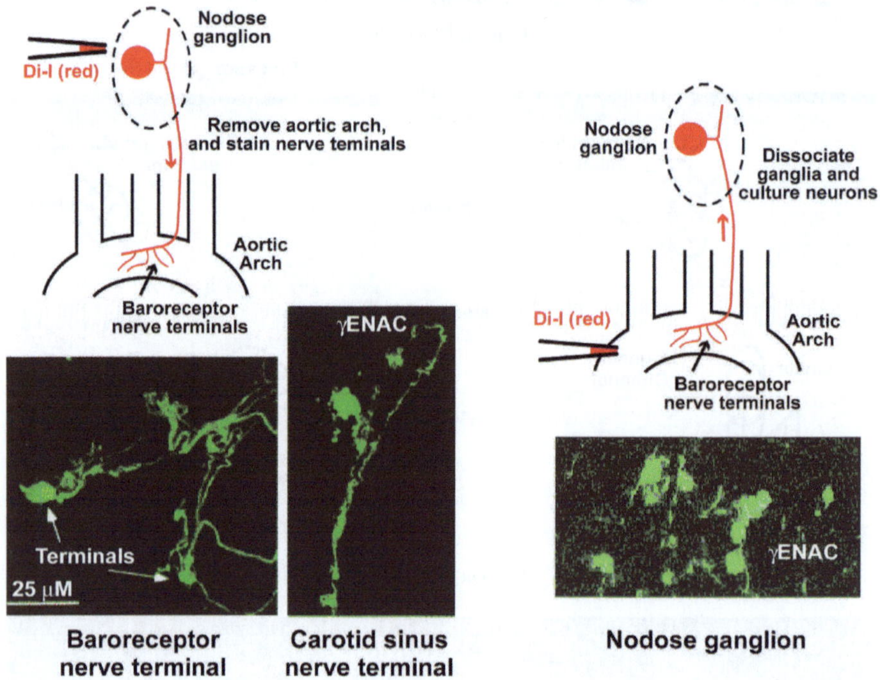

Color Plate 10.

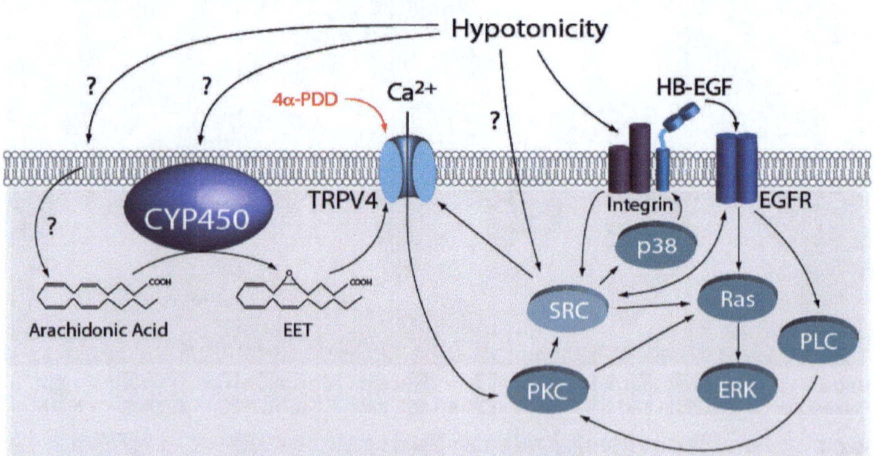

Color Plate 11.

Part III
Other Ion Channels and Biosensors

9
Ion Channels in Shear Stress Sensing in Vascular Endothelium

Ion Channels in Vascular Mechanotransduction

Abdul I. Barakat*, Deborah K. Lieu, and Andrea Gojova

Abstract: Endothelial cell (EC) responsiveness to fluid-mechanical shear stress is essential for normal vascular function and may play a role in the localization of early atherosclerotic lesions. Although ECs are known to be exquisitely sensitive to flow, the precise mechanisms by which ECs sense and respond to shear stress remain incompletely understood. The activation of flow-sensitive ion channels is one of the most rapid endothelial responses to shear stress; therefore, these ion channels have been proposed as candidate flow sensors. A central role for flow-sensitive ion channels in EC shear sensing is supported by recent data demonstrating that blocking these ion channels profoundly affects downstream endothelial flow signaling.

This chapter briefly describes the shear stress environment within medium and large arteries and highlights the wide array of EC responses induced by shear stress. Specific attention is focused on the differential responsiveness of ECs to different types of shear stress. The impact of shear stress on EC ion channels is described in detail, and a potential role for flow-sensitive ion channels in endowing ECs with the ability to distinguish among different types of flow is proposed. The chapter also addresses the implications of ion channel activation for overall endothelial responsiveness to flow, the potential mechanisms by which shear stress activates ion channels, and the remaining gaps in our understanding of shear stress-sensitive ion channels. Finally, suggestions are made for future research directions in this field.

9.1. Arterial Flow and Mechanotransduction in Vascular Endothelium

9.1.1. Central Importance of Endothelial Flow Sensing and Responsiveness

The vascular endothelium, the innermost layer of cells lining the inner surfaces of all blood vessels, is capable of sensing and transducing mechanical forces

* Mechanical and Aeronautical Engineering, University of California, Davis, One Shields Avenue, Davis, CA 95616, abarakat@ucdavis.edu

associated with blood flow. In fact, mechanosensing and mechanotransduction in endothelium are essential for normal vascular functioning including vasoregulation in response to acute changes in blood flow and arterial wall remodeling following chronic hemodynamic alterations.[1,2] Endothelial responses to flow also appear to play a central role in the development of atherosclerosis. Early atherosclerotic lesions develop in medium and large arteries and localize preferentially at arterial branchings and bifurcations where the flow field is multidirectional and highly "disturbed."[3–5] Recent studies suggest that "disturbed" flow might induce endothelial cell (EC) dysfunction, with consequent deleterious implications to atherogenesis. The nature of this flow disturbance will be discussed in more detail in the next section of this chapter.

9.1.2. Mechanical Stress Field in Medium and Large Arteries

The vascular endothelium is constantly exposed to mechanical stresses due to blood flow. In general, blood flow exposes ECs to a mechanical stress tensor with various components. Because arteries are pressure vessels, one component of the stress tensor is due to the transmural pressure difference and leads to a circumferential (or hoop) stress with resultant stretch forces on the cells within the arterial wall. The other components of the stress field to which ECs are constantly subjected include normal pressure forces and tangential shear (or frictional) stresses due to viscous blood flow. In the present chapter, we will focus exclusively on shear stress and its impact on endothelium.

Fluid-mechanical shear stress on a surface is defined as the frictional force on the surface per unit area. As depicted in Fig. 9.1, shear stress on a surface is mathematically defined as the product of the dynamic viscosity of the fluid and the gradient of the velocity vector evaluated at the surface. Therefore, increased near-surface flow velocities translate into higher levels of shear to which the surface is subjected. Within medium and large arteries, blood flow is pulsatile; thus, the endothelial surface is exposed to a highly dynamic shear stress environment. In relatively straight arterial regions away from branches and bifurcations, blood velocity is primarily axial without considerable directional reversal during the course of the cardiac cycle. Hence, ECs at these arterial sites are exposed to nonreversing

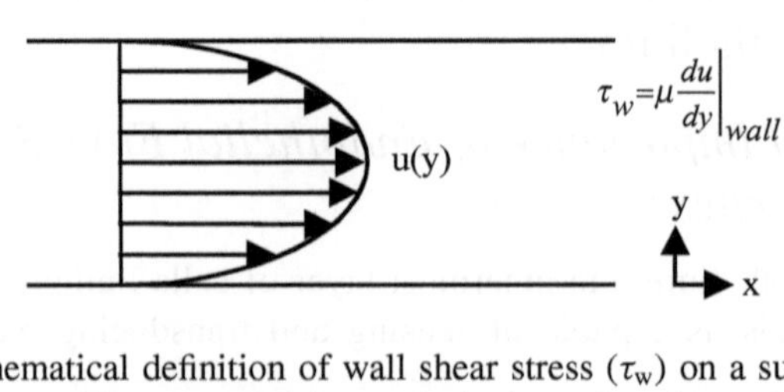

FIGURE 9.1. Mathematical definition of wall shear stress (τ_w) on a surface. The velocity profile within the vessel is assumed to be purely in the x-direction and is denoted by $u(y)$. The dynamic viscosity of the fluid is denoted by μ.

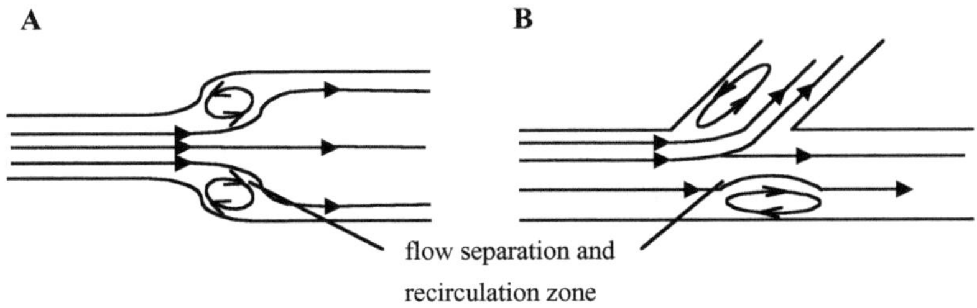

FIGURE 9.2. Examples of flow separation and recirculation in arterial geometries. (A) Arterial expansion. (B) Arterial branching. The extent and intensity of separation and recirculation depend on the detailed geometry and flow conditions.

pulsatile flow. For these vascular locations, experimental and computational studies have established that the mean shear stress on the endothelial surface during the cardiac cycle is ~10–20 dyne/cm^2 (1–2 Pa).[6,7] Blood flow pulsatility leads to excursions about this mean value with a frequency of ~1 Hz. These excursions lead to temporal gradients of shear stress on the EC surface during the cardiac cycle.

In regions of arterial curvature and in the vicinity of branches and bifurcations, the flow field is considerably more complex. Fluid streamline curvature within these regions leads to significant secondary flow motion and thus to a multidirectional shear stress profile with both axial and circumferential components. Secondary flows, when combined with the primary axial flows, often lead to forward-moving helical flow motion. Recent evidence suggests that helical flows act to circumferentially homogenize the wall shear stress distribution so that ECs exposed to such flows are not subjected to extrema in shear stress.[8] Another phenomenon that occurs in regions of rapid changes in arterial geometry is flow separation from the arterial wall as highly inertial fluid fails to remain attached to the arterial wall as it negotiates sharp geometric changes (Fig. 9.2). Regions of flow separation are often occupied by recirculating flow zones within which wall shear stresses are typically very low (~1–5 dyne/cm^2).

Indeed, it is this flow separation and recirculation behavior that is most commonly referred to as the type of "disturbed" flow that appears to best correlate with the localization of early atherosclerotic lesions. Because the wall shear stresses within regions of flow disturbance are lower than those in zones of undisturbed flow, the endothelium is subjected to significant spatial gradients in shear stress. Another particularly important feature of flow separation and recirculation zones is their periodic appearance and disappearance during the course of the pulsatile cardiac cycle. This highly dynamic situation results in ECs within disturbed flow zones being exposed to oscillatory flow. As will be detailed in a subsequent section, there is mounting evidence that oscillatory flow induces a form of EC dysfunction that may be involved in the development of atherosclerosis.

9.2. Endothelial Responses to Shear Stress

Research over the past two decades has established definitively that shear stress elicits a wide array of humoral, metabolic, and structural responses in ECs. Because there are a number of excellent reviews on this topic,[9–15] we will only provide a broad overview of endothelial responsiveness to shear stress and its implications to overall vascular function and pathology.

EC responses to steady (time-independent) shear stress exhibit a wide range of time constants. The fastest responses occur virtually immediately upon flow onset and include activation of flow-sensitive ion channels,[9,16–18] changes in cell membrane fluidity,[19,20] induction of GTP-binding proteins (G proteins),[21,22] and alterations in intracellular pH.[23] Because of their very rapid nature and their proximity to the cell membrane, these responses are thought to be involved in shear stress sensing. The immediate responses are rapidly followed by the stimulation of a number of intracellular signaling events that occur over a period of a few minutes and include mobilization of intracellular calcium,[24–27] activation of integrins,[28,29] induction of mitogen-activated protein (MAP) kinase signaling,[30,31] and phosphorylation of the serine/threonine kinase Akt.[32] These responses are viewed as biochemical signaling events within the shear stress transduction cascade. Another particularly intriguing rapid response is the deformation of intermediate filaments,[33] which has been hypothesized to constitute a form of direct mechanical signaling pathway that serves to transmit the flow signal from the cell surface to various intracellular sites. The notion of a hard-wired EC where cytoskeletal elements enable direct transmission of a mechanical signal to intracellular transduction sites is viewed as a signaling mechanism that is complementary to the more fully characterized biochemical signaling pathways.

Shear stress sensing and initial intracellular signaling are rapidly followed by activation of a number of transcription factors including nuclear factor activator protein-1 (AP-1) and nuclear factor kappa-B (NFκB).[34] Subsequently, various genetic alterations ensue, beginning with the induction of the proto-oncogenes *c-fos, c-myc,* and *c-jun*[35,36] and proceeding to the transcriptional changes in a variety of shear stress-responsive genes.[12,14,37,38]

The diversity of the genetic responses to flow is likely an indication of the central importance of flow signaling in regulating EC function under both normal and pathological conditions. Shear stress regulates the expression of vasoactive and growth factors including endothelin-1 (ET-1),[39,40] nitric oxide synthase (eNOS),[41–43] and platelet-derived growth factor A and B chains (PDGF-A and -B).[44–46] Shear stress also regulates the expression of both pro- and anti-inflammatory genes including the adhesion molecules intracellular adhesion molecule-1 (ICAM-1) and vascular cell adhesion molecule-1 (VCAM-1) and transforming growth factor beta-1 (TGF-β1).[47–50] The flow-induced genetic alterations eventually lead to metabolic and structural changes that include altered macromolecular uptake and internalization,[51,52] extensive remodeling of cytoskeleton,[53–55] and cellular elongation and alignment in the direction of the applied shear stress.[53,56,57]

A number of recent studies have revealed that beyond being responsive to shear stress, ECs respond differently to different types of shear stress. This is particularly relevant in light of the preferential localization of early atherosclerotic lesions within regions of oscillatory flow as already described. Indeed, oscillatory shear stress elicits dramatically different endothelial responses than either steady or nonreversing pulsatile shear stress. For example, while both steady and nonreversing pulsatile shear stress mobilize intracellular calcium oscillations and induce extensive cytoskeletal and morphological remodeling in ECs, oscillatory shear stress does not elicit either response.[47,58,59] Steady and oscillatory shear stress also appear to have different effects on the expression of a number of flow-responsive genes and proteins.[48,60,61]

How do ECs discriminate among different types of shear stress? How does endothelial differential responsiveness to different types of shear stress occur? It appears logical to postulate that this differential responsiveness requires the cells to be capable of sensing and resolving the different components of the shear stress waveform including the amplitude and frequency. Here, we present evidence in support of a central role for flow-sensitive ion channels in shear stress sensing and responsiveness. We also discuss recent results from our laboratory that suggest that these ion channels may also endow ECs with the capability to distinguish among different types of shear stress.

9.3. Effect of Shear Stress on ion Channels in ECs

Several excellent reviews have described the many different types of ion channels in vascular ECs.[62,63] We will focus exclusively on ion channels that are activated by shear stress. To date, several different types of shear stress-sensitive ion channels have been documented in ECs; however, the exact mechanism of activation of these channels remains unknown. Therefore, our discussion of shear-activated ion channels will include channels that are activated either "directly" or "indirectly" by shear stress.

9.3.1. K^+ Channels

Inwardly rectifying K^+ (IRK) channels dominate the resting membrane potential of vascular ECs; therefore, it is not surprising that an IRK channel was the type of ion channel first reported to be responsive to shear stress.[64] Using whole-cell patch clamping, Olesen et al. determined that bovine aortic ECs in culture express flow-sensitive IRK channels and that the extent of channel activation depended on the magnitude of applied shear stress. The current initiated at a shear stress as low as 0.08 dyne/cm^2, reached half-maximal activation at ~0.7 dyne/cm^2, and saturated above ~10 dyne/cm^2. As is typical of IRK channels, the current was entirely inactivated by either Ba^{2+} or Cs^+. Around the same time, Nakache and Gaub used membrane potential-sensitive fluorescent dyes to establish that steady shear stresses up to 120 dyne/cm^2 induced progressively increasing levels of membrane hyperpolarization in bovine pulmonary artery ECs, consistent with the activation of K^+ channels.[17] Subsequent to these initial reports, single channel recordings demonstrated that shear stress-sensitive IRK channels in bovine aortic ECs have a unitary conductance of ~30 pS.[16] More recently, flow-sensitive IRK channels have been reported to be equally responsive to steady shear stress as they are to oscillatory shear stress with a frequency in the range 0.2 to 1 Hz.[65] Interestingly, these channels are not responsive to oscillatory flow with a frequency of 5 Hz.[65] The implications of the sensitivity of K^+ channels to the frequency of an oscillatory flow signal will be discussed in more detail in a subsequent section.

Information on the structure of shear-sensitive K^+ channels remains limited. Analysis of an IRK channel cloned from bovine aortic ECs and whose mRNA expression increases with shear stress revealed that it belongs to the Kir2.1 family.[66] Expression of this ion channel in *Xenopus* oocytes preserves its inward-rectifying behavior.[66] More recently, the same group has demonstrated that when this cloned channel is expressed in oocytes, its sensitivity to flow sensitivity is preserved.[67] Interestingly, flow must be sustained for a period of at least 2 sec for the current to appear, the current does not exhibit desensitization during flow, and current inactivation is not complete until 1-5 min after flow cessation.[67] Finally, while the size of the current is dependent on the shear stress level as previously described by,[64] the rate of activation appears to be independent of the amplitude of applied shear stress.

In addition to IRK channels, Ca^{2+}-sensitive K^+ channels (K_{Ca}) have also been found to be responsive to flow in bovine coronary ECs,[68] human umbilical vein ECs,[69] and human aortic ECs but not in human capillary ECs.[70] K_{Ca}-mediated hyperpolarization of the EC membrane has been shown using membrane potential dyes to be induced by pulsatile but not by steady shear stress.[68] A steady shear stress of either 15 dyne/cm^2 applied for as little as 2 hr or 5 dyne/cm^2 applied for a period of 24 hr significantly increase the expression of K_{Ca}, while a shear stress of 0.5 dyne/cm^2 fails to elicit the transcriptional response.[69]

From a more functional standpoint, whole-cell measurements show that the shear stress-induced K_{Ca} current is blocked by MEK1/2 or p38 inhibitors, suggesting dependence of the current on the Ras/Raf/MEK/ERK signaling system.[69] Another type of K^+ channels whose expression appears to be regulated by shear stress is the ATP-sensitive K^+ channel (K_{ATP}). In both rat pulmonary microvascular and bovine pulmonary artery ECs, K_{ATP} expression is increased significantly following exposure of the cells to shear stresses higher than 5 dyne/cm^2 for a period of 24 hr.[71] Furthermore, flow appears to directly impact EC membrane potential through an effect on K_{ATP} channels.[71]

The exact mechanisms governing the activation of K^+ channels by shear stress are unknown. Shear-induced IRK channel activation does not appear to be affected by osmotic stress, intracellular calcium levels, microfilament or microtubule organization, phospholipase C, inositol phosphate, or protein kinase G; however, IRK activation is abolished by tyrosine kinase inhibition.[67] IRK activation may also be affected by membrane tension or fluidity. This is supported by the observation that enriching the EC membrane with cholesterol, which stiffens the cell membrane, attenuates IRK current density by affecting the extent of ion channel activation.[72] K_{ATP} sensitivity to shear stress has been hypothesized to involve cytoskeletal components or cytoskeleton-associated structures including focal adhesion kinases, integrins, and caveolae.[71]

9.3.2. Cl^- Channels

The discovery of shear stress-sensitive Cl^- channels in ECs came much later than the discovery of flow-sensitive K^+ channels. Several groups in the same year published data on Cl^-channel activation by shear stress in bovine and human aortic ECs but not in human capillary ECs.[18,70,73] Significantly, Cl^- channels are activated by shear stress independently of other ion channels, and activation of these channels leads to cell membrane depolarization following the initial hyperpolarization that is mediated by the activation of shear stress-sensitive K^+ channels.[73] The fact that hyperpolarization precedes depolarization in spite of the higher driving potential for Cl^- than K^+ ions suggests that flow-sensitive Cl^- channels are activated less rapidly than flow-sensitive K^+ channels.[65,73] In bovine aortic ECs, flow-sensitive Cl^- channels are robustly responsive to steady shear stress but are virtually insensitive to oscillatory flow.[65] In bovine coronary ECs, Cl^- channels appear to be activated by both steady and nonreversing pulsatile shear stress.[68]

The class of Cl^- channels to which flow-sensitive Cl^- channels belong remains to be established. It is even possible that different types of flow-sensitive Cl^- channels are expressed in different types of ECs. The time dependence and outward-rectifying behavior of flow-sensitive Cl^- channels resembles that of Ca^{2+}-activated Cl^- channels.[70] Volume-regulated anion channels (VRAC) have also been suggested as a possible type of flow-sensitive Cl^- channels. VRAC exhibit an interesting behavior in response to shear, although the response is only triggered in the presence of an osmotic stress. In the presence of osmotic stress, an increase in shear stress from 0.1 to 1 dyne/cm^2 elicits significant and stable VRAC activation, while at a shear stress of 3 dyne/cm^2, the activation is transient. At higher shear stresses (5–20 dyne/cm^2), the activation is abolished.[74] In calf pulmonary artery ECs, VRAC are regulated by caveolin-1.[75] In bovine aortic ECs, VRAC activity is suppressed by an increase in membrane cholesterol, which rigidifies the membrane, possibly by increasing the membrane deformation energy needed to activate the ion channels.[76]

9.3.3. Ca^{2+} Channels

In response to shear stress, intracellular Ca^{2+} concentration appears to increase both due to release from intracellular stores as well as influx from the extracellular space. In human umbilical vein ECs, shear stress activates a cation channel that is more permeable to Ca^{2+} than to other cations,[77] and Ca^{2+} channels appear to be sensitive to a shear stress as low as 5 dyne/cm^2.[78] Flow has been shown to activate a nonselective cation channel that allows influx of Ca^{2+} in human aortic

ECs but not in human capillary ECs.[70] In human aortic ECs, the intracellular Ca^{2+} increase triggered by shear stress depends on the extracellular Ca^{2+} concentration and appears to occur through nonselective cation channels.[18] Although membrane potential modulates Ca^{2+} influx, changes in membrane potential alone without flow do not change intracellular Ca^{2+} concentration.[18]

9.3.4. Na^+ Channels

Less is known about Na^+ channels than other flow-sensitive ion channels. Initial evidence for the involvement of Na channels in flow signaling was indirect and consisted of amplification of shear stress-induced ERK1/2 when Na influx was abolished.[79] The Na^+ channels involved showed homology with voltage-gated Na^+ channels SCN8a and SCN4a.[79] Flow-activated Na^+ channels were subsequently demonstrated directly in rat cardiac microvascular ECs using patch clamping.[80] In human umbilical vein ECs, Na^+ can enter the cells through shear stress-sensitive cation channels that though more permeable to Ca^{2+}, are also permeable to Na.[77] Interestingly, flow-sensitive Na^+ channels appear to be absent in bovine aortic ECs.[73]

9.4. Implications of Flow-Sensitive Ion Channel Activation for Overall Flow Signaling

As already described, endothelial responsiveness to shear stress is essential for normal vascular function and plays a role in the development of atherosclerosis. Although the role of flow-sensitive ion channels in regulating flow signaling in ECs remains to be established, these ion channels are likely to have profound implications.

The most direct impact of ion channel activation is a change in EC membrane potential. In nonexcitable cells, changes in the membrane potential affect the driving force for Ca^{2+}, which is an important second messenger. Indeed, activation of K^+ channels and changes in membrane voltage have been shown to greatly impact Ca^{2+} influx from the extracellular space into rat thoracic aortic ECs.[81] However, in addition to its effect on membrane potential, activation of flow-sensitive ion channels appears to have a number of other effects. Ion channel activation likely impacts intracellular pH, which may subsequently alter enzymatic activity within ECs. Indeed, an increase in shear stress from 0.3 to 13.4 dyne/cm^2 has been shown to cause immediate acidification of the EC cytosol,[23] while a decrease in shear stress leads to more alkaline pH levels.[82] Interestingly, pH affects activation of VRAC through direct proton binding to the ion channel, resulting in a modest increase in the current following mild intracellular acidification and a considerably larger decrease subsequent to substantial intracellular acidification or alkalization.[83]

The activity of flow-sensitive ion channels also appears to impact various aspects of downstream flow signaling in ECs. Pharmacological blockers of flow-sensitive

K^+ channels considerably attenuate several shear stress-induced endothelial responses including the release of an endogenous nitrovasodilator,[84] the upregulation of TGF-β1 mRNA expression,[49] and the down-regulation of ET-1.[12] More recently, blocking shear stress-sensitive K^+ and Cl^-channels has been shown to dramatically reduce flow-induced increases in endothelial Na-K-Cl cotransport protein activity,[61] while flow-sensitive Cl^- channels but not K^+ channels appear to be involved in regulating Akt phosphorylation by flow (Gautam et al., in press).

In addition to a potentially central role in overall endothelial responsiveness to flow, recent data raise the intriguing possibility that shear stress-sensitive ion channels may endow ECs with the ability to distinguish among and respond differently to different types of flow.[65] While steady shear stress activates both K^+ and Cl^- channels in bovine aortic ECs, 0.2- or 1-Hz oscillatory shear stress fully activates K^+ while activating Cl^- channels to a much smaller degree. 5-Hz oscillatory flow activates neither K^+ nor Cl^- channels. It has been hypothesized that flow-sensitive K^+ and Cl^- channels may constitute components of a mechanosensory system where the relative activation of the two types of channels (and its impact on both membrane potential and intracellular K^+ and Cl^- concentration) may provide a signal that allows ECs to distinguish between steady and oscillatory flow as well as among different frequencies of flow oscillation.

9.5. Mechanisms of Ion Channel Activation by Shear Stress

The mechanisms by which stretch-activated ion channels are activated have been studied in detail.[85] In contrast, the mechanisms governing the activation of shear stress-sensitive ion channels remain to be elucidated. One possible pathway is channel activation through a direct physical effect of the shear force on extracellular domains of the channel protein. One difficulty with this notion, however, is that the size of these domains is expected to be very small (~10 nm), so that the energy imparted by the flow would be expected to be too small to lead to significant deformations or conformational changes.

Another mechanism that has been hypothesized for activation of shear stress-sensitive ion channels is the transmission of the flow force to the channels via the cellular cytoskeleton. Within this construct, flow leads to an alteration in intracellular tension, which is accommodated by deformation of cytoskeletal elements. This deformation would then control gating of cytoskeleton-associated ion channels. Support for this model is provided by several observations. Firstly, there is considerable evidence that force transmission in eukaryotic cells generally occurs through the cytoskeleton.[85] Secondly, ion channels in certain cell types have also been shown to be directly coupled to membrane cytoskeleton via ankyrin.[85] Thirdly, the elastic constant of a patch of cell membrane is ~50 dyne/cm, at least ten times smaller than the elastic constant of a pure lipid bilayer,[86] suggesting that the force activating ion channels is sustained primarily by the cytoskeleton.[85] Finally, disruption of cytoskeleton has been shown to modulate ion channel activity in *Xenopus* oocytes,[87] leukocytes,[88] Jurkat cells,[88] and rat colon sensory neurons.[89]

Mechanosensitive ion channels in bacteria have been shown to respond to changes in membrane tension.[85] As already described, exposure of ECs to shear stress leads to a rapid increase in cell membrane fluidity.[19,20] It is possible that the flow-induced activation of ion channels occurs following the flow-induced increase in membrane fluidity, which may alter cell membrane tension. Indeed, increasing membrane fluidity by depletion of cholesterol from the EC membrane enhances the development of VRAC, while enriching the membrane with cholesterol has the opposite effect.[76,90] Therefore, the possibility remains that activation of shear stress-sensitive ion channels may occur indirectly and occurs through the impact of the shear force on cell membrane tension. In addition to the hypothesized mechanisms outlined above, the possibility that shear stress-sensitive ion channels are ligand-gated cannot be excluded. This notion is supported by data suggesting that shear stress-sensitive K^+ and Cl^- channels in ECs are modulated by pertussis toxin-sensitive G proteins whose activation may elicit production of a number of second messengers including cAMP, IP3, DAG, Ca^{2+}, and arachidonic acid.[91]

9.6. Gaps in Present Understanding and Future Directions

While much has been learned over the past 15 years about shear stress-sensitive ion channels in vascular ECs, there are many questions that remain to be answered. It is essential to systematically establish which types of flow-sensitive ion channels are present in which types of ECs. Furthermore, flow-induced ion currents have to date only been studied in response to short episodes of flow stimulation. *In vivo*, ECs are exposed to flow chronically. In the presence of sustained flow, shear stress-sensitive ion channels might be expected to exhibit desensitization. Therefore, it is essential to extend many of the studies to longer flow periods in order to establish the kinetics for desensitization of the various types of flow-sensitive ion channels. Performing long-term flow studies, however, is greatly complicated by the fact that recording techniques, most notably patch-clamping protocols, do not permit continuous acquisition for extended periods of time. Therefore, development of improved ion current measurement technology that permits long term channel monitoring under flow would be greatly helpful.

Recent studies have revealed that different types of flow may not have the same impact on the different types of shear stress-sensitive ion channels. Given the complexity of arterial flow fields and the range of shear stress levels and characteristics that are encountered *in vivo*, it is important to further characterize the responsiveness of the different types of flow-sensitive ion channels to different levels and types of shear stress. Such information would provide critical insight into which ion channels would be expected to be activated within specific arterial regions.

In this chapter, we have proposed that ion channels play a central role in shear stress sensing and responsiveness. It is important to directly test this notion to the extent possible. This may be accomplished by probing the effect of alterations in ion channel activity on various flow-stimulated endothelial signaling events including mobilization of intracellular Ca^{2+}, activation of MAP kinase, and phosphorylation of Akt. One complication in this regard is that interfering with ion channel activity is currently often confined to the use of pharmacological agents. Therefore, there is great need for cloning and characterization of the various types of shear stress-sensitive ion channels in ECs in order to enable development of dominant negative or positive constructs.

Finally, all studies to date have been performed on single ECs or monolayers of ECs in culture. *In vivo*, arterial ECs are in constant communication with other vascular cell types, most notably smooth muscle cells. It will be important to perform ion channel studies on ECs co-cultured with smooth muscle cells in order to establish whether or not EC-smooth muscle cell communication impacts the activation of ion channels by shear stress.

Acknowledgments. This work was supported in part by grants from The Whitaker Foundation, Pfizer/Parke-Davis, and Philip Morris USA.

References

1. Langille, B.L. and O'Donnell, F., 1986, Reductions in arterial diameter produced by chronic decreases in blood flow are endothelium-dependent, *Science* **231**:405–7.
2. Pohl, U., Holtz, J., Busse, R. and Bassenge, E., 1986, Crucial role of endothelium in the vasodilator response to increased flow in vivo, *Hypertension* **8**:37–44.
3. Caro, C.G., Fitz-Gerald, J.M. and Schroter, R.C., 1969, Arterial wall shear and distribution of early atheroma in man, *Nature* **223**:1159–60.
4. Nerem, R.M., 1992, Vascular fluid mechanics, the arterial wall, and atherosclerosis, *J. Biomech. Eng.* **114**:274–82.
5. Svindland, A. and Walloe, L., 1985, Distribution pattern of sudanophilic plaques in the descending thoracic and proximal abdominal human aorta, *Atherosclerosis* **57**:219–24.
6. Asakura, T., and Karino, T., 1990, Flow patterns and spatial distribution of atherosclerotic lesions in human coronary arteries, *Circ. Res.* **66**:1045–66.
7. Barakat, A.I., Karino, T., and Colton, C.K., 1997, Microcinematographic studies of flow patterns in the excised rabbit aorta and its major branches, *Biorheology* **34**:195–221.
8. Sherwin, S.J., Shah, O., Doorly, D.J., Peiro, J., Papaharilaou, Y., Watkins, N., Caro, C.G. and Dumoulin, C.L., 2000, The influence of out-of-plane geometry on the flow within a distal end-to-side anastomosis, *J. Biomech. Eng.* **122**:86–95.
9. Barakat, A.I., 1999, Responsiveness of vascular endothelium to shear stress: potential role of ion channels and cellular cytoskeleton (review), *Int. J. Mol. Med.* **4**:323–32.
10. Davies, P.F., 1995, Flow-mediated endothelial mechanotransduction, *Physiol. Rev.* **75**:519–560.
11. Fisher, A.B., Chien, S., Barakat, A.I. and Nerem, R.M., 2001, Endothelial cellular response to altered shear stress, *Am. J. Physiol. Lung Cell Mol. Physiol.* **281**:L529–33.
12. Malek, A.M. and Izumo, S., 1994, Molecular aspects of signal transduction of shear stress in the endothelial cell, *J. Hypertens.* **12**:989–99.
13. Papadaki, M. and Eskin, S.G., 1997, Effects of fluid shear stress on gene regulation of vascular cells, *Biotechnol. Prog.* **13**:209–21.

14. Resnick, N. and Gimbrone, M.A., Jr., 1995, Hemodynamic forces are complex regulators of endothelial gene expression, *Faseb J* **9**:874–82.
15. Traub, O. and Berk, B.C., 1998, Laminar shear stress: mechanisms by which endothelial cells transduce an atheroprotective force, *Arterioscler. Thromb. Vasc. Biol.* **18**:677–85.
16. Jacobs, E.R., Cheliakine, C., Gebremedhin, D., Birks, E.K., Davies, P.F. and Harder, D.R., 1995, Shear activated channels in cell-attached patches of cultured bovine aortic endothelial cells, *Pflugers Archiv. Eur. J. Physiol.* **431**:129–31.
17. Nakache, M. and Gaub, H.E., 1988, Hydrodynamic hyperpolarization of endothelial cells, *Proc. Natl. Acad. Sci. USA* **85**:1841–1843.
18. Nakao, M., Ono, K., Fujisawa, S. and Iijima, T., 1999, Mechanical stress-induced Ca2+ entry and Cl- current in cultured human aortic endothelial cells, *Am. J. Physiol.* **276**:C238–C249.
19. Butler, P.J., Norwich, G., Weinbaum, S. and Chien, S., 2001, Shear stress induces a time- and position-dependent increase in endothelial cell membrane fluidity, *Am. J. Physiol.* **280**:C962–C969.
20. Haidekker, M.A., L'Heureux, N. and Frangos, J.A., 2000, Fluid shear stress increases membrane fluidity in endothelial cells: a study with DCVJ fluorescence, *Am. J. Physiol. Heart Circ. Physiol.* **278**:H1401–6.
21. Gudi, S., Nolan, J.P. and Frangos, J.A., 1998, Modulation of GTPase activity of G proteins by fluid shear stress and phospholipid composition, *Proc. Natl. Acad. Sci. USA* **95**:2515–9.
22. Gudi, S.R., Clark, C.B. and Frangos, J.A., 1996, Fluid flow rapidly activates G proteins in human endothelial cells. Involvement of G proteins in mechanochemical signal transduction, *Circ. Res.* **79**:834–9.
23. Ziegelstein, R.C., Cheng, L. and Capogrossi, M.C., 1992, Flow-dependent cytosolic acidification of vascular endothelial cells, *Science* **258**:656–9.
24. Ando, J., Komatsuda, T., and Kamiya, A., 1988, Cytoplasmic calcium response to fluid shear stress in cultured vascular endothelial cells, *In Vitro Cell Dev. Biol.* **24**: 871–7.
25. Dull, R.O. and Davies, P.F., 1991, Flow modulation of agonist (ATP)-response (Ca2+) coupling in vascular endothelial cells, *Am. J. Physiol.* **261**:H149–54.
26. Geiger, R.V., Berk, B.C., Alexander, R.W. and Nerem, R.M., 1992, Flow-induced calcium transients in single endothelial cells: spatial and temporal analysis, *Am. J. Physio.l* **262**:C1411–7.
27. Shen, J., Luscinskas, F.W., Connolly, A., Dewey, C.F., Jr. and Gimbrone, M.A., Jr., 1992, Fluid shear stress modulates cytosolic free calcium in vascular endothelial cells, *Am. J. Physiol.* **262**:C384–90.
28. Shyy, J.Y. and Chien, S., 1997, Role of integrins in cellular responses to mechanical stress and adhesion, *Curr. Opin. Cell Biol.* **9**:707–13.
29. Tzima, E., del Pozo, M.A., Shattil, S.J., Chien, S. and Schwartz, M.A., 2001, Activation of integrins in endothelial cells by fluid shear stress mediates Rho-dependent cytoskeletal alignment, *Embo. J.* **20**:4639–47.
30. Tseng, H., Peterson, T.E. and Berk, B.C., 1995, Fluid shear stress stimulates mitogen-activated protein kinase in endothelial cells, *Circ. Res.* **77**:869–78.
31. Yan, C., Takahashi, M., Okuda, M., Lee, J.D. and Berk, B.C., 1999, Fluid shear stress stimulates big mitogen-activated protein kinase 1 (BMK1) activity in endothelial cells. Dependence on tyrosine kinases and intracellular calcium, *J. Biol. Chem.* **274**:143–50.

32. Dimmeler, S., Assmus, B., Hermann, C., Haendeler, J. and Zeiher, A.M., 1998, Fluid shear stress stimulates phosphorylation of Akt in human endothelial cells: involvement in suppression of apoptosis, *Circ. Res.* **83**:334–41.
33. Helmke, B.P., Goldman, R.D. and Davies, P.F., 2000, Rapid displacement of vimentin intermediate filaments in living endothelial cells exposed to flow, *Circ. Res.* **86**:745–52.
34. Lan, Q., Mercurius, K.O. and Davies, P.F., 1994, Stimulation of transcription factors NF kappa B and AP1 in endothelial cells subjected to shear stress, *Biochem. Biophys. Res. Commun.* **201**:950–6.
35. Hsieh, H.J., Li, N.Q. and Frangos, J.A., 1993, Pulsatile and steady flow induces c-fos expression in human endothelial cells, *J. Cell Physiol.* **154**:143–51.
36. Ranjan, V. and Diamond, S.L., 1993, Fluid shear stress induces synthesis and nuclear localization of c-fos in cultured human endothelial cells, *Biochem. Biophys. Res. Commun.* **196**:79–84.
37. Braddock, M., Schwachtgen, J.L., Houston, P., Dickson, M.C., Lee, M.J. and Campbell, C.J., 1998, Fluid shear stress modulation of gene expression in endothelial cells, *News Physiol. Sci.* **13**:241–246.
38. Garcia-Cardena, G., Comander, J., Anderson, K.R., Blackman, B.R. and Gimbrone, M.A., Jr., 2001, Biomechanical activation of vascular endothelium as a determinant of its functional phenotype, *Proc. Natl. Acad. Sci. USA* **98**:4478–85.
39. Malek, A.M., Gibbons, G.H., Dzau, V.J. and Izumo, S., 1993, Fluid shear stress differentially modulates expression of genes encoding basic fibroblast growth factor and platelet-derived growth factor B chain in vascular endothelium, *J. Clin. Invest.* **92**:2013–21.
40. Yoshizumi, M., Kurihara, H., Sugiyama, T., Takaku, F., Yanagisawa, M., Masaki, T. and Yazaki, Y., 1989, Hemodynamic shear stress stimulates endothelin production by cultured endothelial cells, *Biochem. Biophys. Res. Commun.* **161**:859–64.
41. Malek, A.M., Izumo, S. and Alper, S.L., 1999, Modulation by pathophysiological stimuli of the shear stress-induced up-regulation of endothelial nitric oxide synthase expression in endothelial cells, *Neurosurgery* **45**:334–44; discussion 344–5.
42. Noris, M., Morigi, M., Donadelli, R., Aiello, S., Foppolo, M., Todeschini, M., Orisio, S., Remuzzi, G. and Remuzzi, A., 1995, Nitric oxide synthesis by cultured endothelial cells is modulated by flow conditions, *Circ. Res.* **76**:536–43.
43. Uematsu, M., Ohara, Y., Navas, J.P., Nishida, K., Murphy, T.J., Alexander, R.W., Nerem, R.M. and Harrison, D.G., 1995, Regulation of endothelial cell nitric oxide synthase mRNA expression by shear stress, *Am. J. Physiol.* **269**:C1371–8.
44. Bao, X., Lu, C., and Frangos, J.A., 1999, Temporal gradient in shear but not steady shear stress induces PDGF-A and MCP-1 expression in endothelial cells: role of NO, NF kappa B, and egr-1, *Arterioscler. Thromb. Vasc. Biol.* **19**:996–1003.
45. Hsieh, H.J., Li, N.Q. and Frangos, J.A., 1991, Shear stress increases endothelial platelet-derived growth factor mRNA levels, *Am. J. Physiol.* **260**:H642–6.
46. Malek, A.M., Greene, A.L. and Izumo, S., 1993, Regulation of endothelin 1 gene by fluid shear stress is transcriptionally mediated and independent of protein kinase C and cAMP, *Proc. Natl. Acad. Sci. USA* **90**:5999–6003.
47. Lum, R.M., Wiley, L.M. and Barakat, A.I., 2000, Influence of different forms of fluid shear stress on vascular endothelial TGF-beta1 mRNA expression, *Int. J. Mol. Med.* **5**:635–41.
48. Nagel, T., Resnick, N., Atkinson, W.J., Dewey, C.F., Jr. and Gimbrone, M.A., Jr., 1994, Shear stress selectively upregulates intercellular adhesion molecule-1 expression in cultured human vascular endothelial cells, *J. Clin. Invest.* **94**:885–91.

49. Ohno, M., Cooke, J.P., Dzau, V.J. and Gibbons, G.H., 1995, Fluid shear stress induces endothelial transforming growth factor beta-1 transcription and production: modulation by potassium channel blockade, *J. Clin. Invest.* **95**:1363–9.
50. Tsuboi, H., Ando, J., Korenaga, R., Takada, Y. and Kamiya, A., 1995, Flow stimulates ICAM-1 expression time and shear stress dependently in cultured human endothelial cells, *Biochem. Biophys. Res. Commun.* **206**:988–96.
51. Kudo, S., Morigaki, R., Saito, J., Ikeda, M., Oka, K. and Tanishita, K., 2000, Shear-stress effect on mitochondrial membrane potential and albumin uptake in cultured endothelial cells, *Biochem. Biophys. Res. Commun.* **270**:616–21.
52. Sprague, E.A., Steinbach, B.L., Nerem, R.M. and Schwartz, C.J., 1987, Influence of a laminar steady-state fluid-imposed wall shear stress on the binding, internalization, and degradation of low-density lipoproteins by cultured arterial endothelium, *Circulation* **76**:648–56.
53. Nerem, R.M., Levesque, M.J. and Cornhill, J.F., 1981, Vascular endothelial morphology as an indicator of the pattern of blood flow, *J. Biomech. Eng.* **103**:172–6.
54. Ookawa, K., Sato, M. and Ohshima, N., 1992, Changes in the microstructure of cultured porcine aortic endothelial cells in the early stage after applying a fluid-imposed shear stress, *J. Biomech.* **25**:1321–8.
55. Wechezak, A.R., Viggers, R.F. and Sauvage, L.R., 1985, Fibronectin and F-actin redistribution in cultured endothelial cells exposed to shear stress, *Lab. Invest.* **53**:639–47.
56. Dewey, C.F., Jr., Bussolari, S.R., Gimbrone, M.A., Jr. and Davies, P.F., 1981, The dynamic response of vascular endothelial cells to fluid shear stress, *J. Biomech. Eng.* **103**:177–85.
57. Eskin, S.G., Ives, C.L., McIntire, L.V. and Navarro, L.T., 1984, Response of cultured endothelial cells to steady flow, *Microvasc. Res.* **28**:87–94.
58. Helmlinger, G., Geiger, R.V., Schreck, S. and Nerem, R.M., 1991, Effects of pulsatile flow on cultured vascular endothelial cell morphology, *J. Biomech. Eng.* **113**:123–31.
59. Helmlinger, G., Berk, B.C. and Nerem, R.M., 1995, Calcium responses of endothelial cell monolayers subjected to pulsatile and steady laminar flow differ, *Am. J. Physiol.* **269**:C367–75.
60. Chappell, D.C., Varner, S.E., Nerem, R.M., Medford, R.M. and Alexander, R.W., 1998, Oscillatory shear stress stimulates adhesion molecule expression in cultured human endothelium, *Circ. Res.* **82**:532–9.
61. Suvatne, J., Barakat, A.I. and O'Donnell, M.E., 2001, Flow-induced expression of endothelial N-K-Cl cotransport: dependence on K+ and Cl− channels, *Am. J. Physiol.* **280**:C216–C227.
62. Nilius, B., Viana, F. and Droogmans, G., 1997, Ion channels in vascular endothelium, *Ann. Rev. Physiol.* **59**:145–170.
63. Nilius, B. and Droogmans, G., 2001, Ion channels and their functional role in vascular endothelium, *Physiol. Rev.* **81**:1415–1459.
64. Olesen, S.P., Clapham, D.E. and Davies, P.F., 1988, Hemodynamic shear-stress activates a K+ current in vascular endothelial cells, *Nature* **331**:168–170.
65. Lieu, D.K., Pappone, P.A. and Barakat, A.I., 2004, Differential membrane potential and ion current responses to different types of shear stress in vascular endothelial cells, *Am. J. Physiol.* **286**:C1367–C1375.
66. Forsyth, S.E., Hoger, A. and Hoger, J.H., 1997, Molecular cloning and expression of a bovine endothelial inward rectifier potassium channel, *Febs. Lett.* **409**:277–282.

67. Hoger, J.H., Ilyin, V.I., Forsyth, S. and Hoger, A., 2002, Shear stress regulates the endothelial Kir2.1 ion channel, *Proc. Natl. Acad. Sci. USA* **99**:7780–7785.
68. Qui, W., Hu, Q., Paolocci, N., Ziegelstein, R.C. and Kass, D.A., 2003, Differential effects of pulsatile versus steady flow on coronary endothelial membrane potential, *Am. J. Physiol.* **285**:H341–H346.
69. Brakemeier, S., Kersten, A., Eichler, I., Grgic, I., Zakrzewicka, A., Hopp, H., Kohler, R. and Hoyer, J., 2003, Shear stress-induced up-regulation of the intermediate-conductance Ca2+-activated K+ channel in human endothelium, *Cardio. Res.* **60**:488–496.
70. Jow, F. and Numann, R., 1999, Fluid flow modulates calcium entry and activates membrane currents in cultured human aortic endothelial cells, *J. Membrane Biol.* **171**:127–139.
71. Chatterjee, S., Al-Mehdi, A., Levitan, I., Stevens, T. and Fisher, A.B., 2003, Shear stress increases expression of a KATP Channel in rat and bovine pulmonary vascular endothelial cells, *Am. J. Physiol. Cell Physiol.* **285**:C959–C967.
72. Romaneneko, V.G., Rothblat, G.H. and Levitan, I., 2002, Modulation of endothelial inward-rectifier K+ current by optical isomers of cholesterol, *Biophys. J.* **83**:3211–3222.
73. Barakat, A.I., Leaver, E.V., Pappone, P.A., and Davies, P.F., 1999, A flow-activated chloride-selective membrane current in vascular endothelial cells, *Circ. Res.* **85**:820–828.
74. Romanenko, V.G., Davies, P.F. and Levitan, I., 2002, Dual effect of fluid shear stress on volume-regulated anion current in bovine aortic endothelial cells, *Am. J. Physiol.* **282**:C708–C718.
75. Trouet, D., Hermans, D., Droogmans, G., Nilius, B. and Eggermont, J., 2001, Inhibition of volume-regulated anion channels by dominant-negative caveolin-1, *Biochem. Biophys. Res. Commun.* **284**:461–465.
76. Levitan, I., Christian, A.E., Tulenko, T.N. and Rothblat, G.H., 2000, Membrane cholesterol content modulates activation of volume-regulated anion current in bovine endothelial cells, *J. Gen. Physiol.* **115**:405–416.
77. Schwarz, G., Droogmans, G. and Nilius, B., 1992, Shear stress induced membrane currents and calcium transients in human vascular endothelial cells, *Pflügers Archiv. Eur. J. Physiol.* **421**:394–396.
78. Brakemeier, S., Eichler, I., Hopp, H., Kohler, R. and Hoyer, J., 2002, Up-regulation of endothelial stretch-activated cation channels by fluid shear stress, *Cardio. Res.* **53**:209–218.
79. Traub, O., T. Ishida, M. Ishida, J.C. Tupper, B.C. Berk, 1999, Shear stress-mediated extracellular signal-regulated kinase activation is regulated by sodium in endothelial cells, *J. Bio.l Chem.* **274**:20144–20150.
80. Moccia, F., Villa, A. and Tanzi, F., 2000, Flow-activated Na+ and K+ current in cardiac microvascular endothelial cells, *J. Mol. Cell. Cardiol.* **32**:1589–1593.
81. Kwan, H., Leung, P., Huang, Y. and Yao, X., 2003, Depletion of intracellular Ca2+ stores sensitizes the flow-induced Ca2+ influx in rat endothelial cells, *Circ. Res.* **92**:286–292.
82. Ziegelstein, R.C., Blank, P.S., Cheng, L. and Capogrossi, M.C., 1998, Cytosolic Alkalinization of vascular endothelial cells produced by an abrupt reduction in fluid shear stress, *Circ. Res.* **82**:803–809.

83. Sabirov, R.Z., Prenen, J., Droogmans, G. and Nilius, B., 2000, Extra- and Intracellular Proton-Binding Sites of Volume-Regulated Anion Channels, *J. Membrane Biol.* **177**:13–22.
84. Cooke, J.P., Rossitch, E., Jr., Andon, N.A., Loscalzo, J. and Dzau, V.J., 1991, Flow activates an endothelial potassium channel to release an endogenous nitrovasodilator, *J. Clin. Invest.* **88**:1663–71.
85. Sachs, F. and Morris, C. (1998) in *Rev. Physio. Biochem. Pharmacol.* (Springer, Berlin), pp. 1–78.
86. Evans, E. and Needham, D., 1987, Physical properties of surfactant bilayer membranes: thermal transitions, elasticity, rigidity, cohesion, and colloidal interactions, *J. Phys. Chem.* **91**:4219–4228.
87. Hamill, O.P. and McBride Jr., D.W., 1997, Induced membrane hypo/hyper-mechanosensitivity: a limitation of patch-clamp recording, *Ann. Rev. Physiol.* **59**:621–31.
88. Downey, G.P., Grinstein, S., Sue, A.Q.A., Czaban, B. and Chan, C.K., 1995, Volume regulation in leukocytes: requirement for an intact cytoskeleton, *J. Cell. Physiol.* **163**:96–104.
89. Su, X., Wachtel, R.E. and Gebhart, G.F., 2000, Mechanosensitive potassium channels in rat colon sensory neurons, *J. Neurophysiol.* **84**:836–43.
90. Romanenko, V.G., Rothblat, G. H. and Levitan, I., 2002, Modulation of Endothelial Inward-Rectifier K+ Current by Optical Isomers of Cholesterol, *Biophys. J.* **83**:3211–3222.
91. Ohno, M., Gibbons, G.H., Dzau, V.J. and Cooke, J.P., 1993, Shear-stress elevates endothelial cGMP: role of a potassium channel and G-protein coupling, *Circulation* **88**:193–197.

10
Redox Signaling in Oxygen Sensing by Vessels

Andrea Olschewski* and E. Kenneth Weir†

10.1. Introduction

Oxidant production and regulation is becoming increasingly important in the study of vascular signaling mechanism. A large number of studies during the last 50 years have provided evidence that vascular preparations show alterations in contractile function over a wide range of O_2 tensions that are observed in physiological systems. Based on observations that reactive oxygen species were vasoactive and appeared to have distinct signaling mechanisms, it was suggested that these species could function in vascular O_2 sensing mechanisms that mediated responses to acute changes in pO_2.[1,2]

Studies in vascular smooth muscle have provided evidence that NAD(P)H oxidases and mitochondrial systems have important roles in generation of reactive oxygen species.[3–5] H_2O_2 appears to be one of the most important oxidant species because it is known to interact with multiple signaling systems. There is substantial evidence that vascular potassium (K) channels are regulated by signaling mechanisms elicited by changes in pO_2.[6] Several redox-related systems have been linked to the control of K channel function suggesting a key role for K channels in the signaling mechanism of pO_2-elicited vascular responses. O_2- and redox-sensitive K channels have been shown to be the effectors in O_2-sensing in several tissues including the renal and pulmonary arteries (PA) and the ductus arteriosus (DA).[6–8]

Although the PA and DA are closely related anatomically, they behave in a diametrically opposite manner to the increase in O_2-tension: the resistance PA dilate while the DA constricts. While all workers in the field would agree on this observation, the mechanism by which the level of oxygen is sensed and transduced into changes in vascular tone remains controversial. In the case of PA and DA smooth muscle cells (SMCs), one sequence of events precipitated by changes in O_2-tension has been well documented. Hypoxic inhibition of one or

*Medical University Graz, Department of Anesthesiology and Intensive Care Medicine, Auen Bruggerplatz 29, A-8036 Graz, Austria, andrea.olschewski@meduni-graz.at
†Department of Medicine, VA Medical Center, Minneapolis, MN 55417

more potassium channels[7,9] leads to membrane depolarization and calcium entry through the voltage-activated calcium channels in the PASMCs.[10–12] The same sequence seems to occur in smooth muscle cells of the DASMCs. but in this case the sequence is initiated by an increase in oxygen tension that inhibits potassium channels, not by hypoxia.[13–15] The effect of the changes in the pO_2 may be mediated through H_2O_2.[8,16] The opposing responses of the pulmonary vascular bed and DA to hypoxia are, at least in part, due to differences in mitochondria function[16] or in the diversity of K channel subtypes.[17] However, in the last 15 years only parts of the puzzle have been completed; new findings are controversial and new questions have arisen. Hopefully, new technologies and investigations in transgenic and gene-deficient mice may clarify the controversies we are currently facing.

10.2. Sources of Reactive Oxygen Species Production

The term *reactive oxygen species* (ROS) describes a group of small, reactive oxygen-containing molecules that are either free-radicals containing oxygen or nitrogen-based unpaired electrons or compounds that are not free radicals themselves, but have oxidizing properties that may contribute to oxidant stress. Among the free radicals, superoxide anion (O_2^-), hydroxyl radical($OH^\bullet$), nitric oxide ($NO^\bullet$) and lipid radicals ($LO^\bullet$, $LOO^\bullet$) are the most prominent. Nonradical ROS, such as hydrogen peroxide (H_2O_2), hypochloric acid (HOCl), and peroxinitrite ($ONOO^-$) are compounds that may be formed under oxidative stress and mediate oxidant signaling to their environment. Enzymes responsible for ROS formation can be (1) constitutively active such as NAD(P)H oxidase, superoxide dismutase (SOD) or cytochrome P450 and the enzyme complexes contributing to the mitochondrial respiratory chain or (2) can be induced under pathologic conditions, such as inducible nitric oxide synthase (iNOS), or xanthine oxidase.

10.2.1. (NAD(P)H) Oxidase

The nicotinamide adenine dinucleotide phosphate reduced (NAD(P)H) oxidases are membrane-associated enzymes that catalyze the 1-electron reduction of oxygen using NADH or NADPH as the electron donor.

$$\mathrm{NAD(P)H} + 2\mathrm{O_2} \rightarrow \mathrm{NAD(P)}^+ + \mathrm{H}^+ + 2\mathrm{O_2^-}$$

Stimulated production of ROS by NAD(P)H oxidase was first described in phagocytic cells such as neutrophils and macrophages related to the transient consumption of oxygen ("the respiratory burst"). During the last decade, production of ROS by NAD(P)H oxidase or by the NADH oxidase isoforms has also been demonstrated in a variety of cells in O_2-sensitive tissues in the O_2 homeostatic system, including neuroepithelial body cells (NEBs),[18] endothelial cells,[19] and carotid bodies.[20]

Several isoforms of this multicomponent enzyme complex have been found in the cardiovascular system. The major sources of ROS in the vessel wall, the vascular NAD(P)H oxidases, are similar in structure to the neutrophil NADPH oxidase, which consists of four major subunits: a cytochrome b558 in the membrane, comprising gp91^{phox} and p22^{phox}, and two cytosolic components, p47^{phox} and p67^{phox}. In most studies, NADH is proposed to be the preferred substrate[21,22], but some investigators find NADPH driven $O_2^{\cdot -}$ generation to predominate.[23,24] The cardiovascular NAD(P)H oxidases are low-output, slow-release enzymes in which the biochemical characteristics differ considerably from those of the neutrophil NAD(P)H oxidase. Estimates of $O_2^{\cdot -}$ production in vascular cells suggest that the capacity of these enzymes is about one-third that of the neutrophil.[25] The kinetics of activation following cell stimulation are also unique; $O_2^{\cdot -}$ is produced in minutes to hours in endothelial cells, vascular smooth muscle cells, and fibroblasts,[22,26,27] in contrast to the almost instantaneous release seen in neutrophils. However, ROS generation drops within seconds in response to hypoxia. Despite these differences, the neutrophil and vascular enzymes share some characteristics; both are inhibited by diphenylene iodonium (DPI), an inhibitor of flavin-containing oxidases, and both are stimulated by agonists and arachidonic acid.[25]

These enzymes have been proposed to be oxygen sensors and to participate in hypoxic vasoconstriction.[21] Wolin et al. reported in bovine pulmonary arteries that H_2O_2 production increased in normoxia, relative to hypoxia, and caused PASMC relaxation through an increase in cGMP[3]. DPI, the nonselective NAD(P)H oxidase inhibitor, would be expected to reduce ROS and thus resemble hypoxia. In isolated and cultured NEB cells DPI does in fact inhibit K^+ current, as would be predicted if H_2O_2 produced by NAD(P)H oxidase usually signals the K^+ current seen in normoxia.[28] The central role of NAD(P)H oxidase has recently been called into question by Archer et al. In this study, the gp91^{phox} knockout mouse had normal hypoxic pulmonary vasoconstriction (HPV) and hypoxia still inhibited K^+ channels in their PASMC,[29] suggesting that NAD(P)H oxidase containing gp91^{phox} is not required for HPV. Another view of the role of NAD(P)H oxidase in O2-sensing comes from Marshall et al. They showed that NAD(P)H oxidase was activated under hypoxic conditions in smooth muscle from small diameter pulmonary arteries.[30] More recently, Weissmann et al. reported the activation of NAD(P)H oxidase during hypoxia with a subsequent increase in H_2O_2.[31] The study of NAD(P)H oxidase homologues may help to settle this controversy.

10.2.2. Superoxide Dismutase (SOD)

SOD-mediates the conversion of $O_2^{\cdot -}$ to H_2O_2. However these two ROS differ greatly in terms of their physiologic properties and biological action. $O_2^{\cdot -}$ can behave as a reducing or an oxidizing agent. It is classically measured by its ability to reduce cytochrome oxidase, whereas H_2O_2 is an oxidant. $O_2^{\cdot -}$ cannot freely penetrate the cell membrane and needs anion channels for passage,[32] while H_2O_2 can easily move through the cell membrane. Superoxide is formed in pulmonary tissue under basal conditions and can be detected from the lung surface and in isolated pulmonary artery rings. In Krebs-perfused rat lungs, superoxide (measured as SOD-sensitive luminol or lucigenin chemiluminescence recorded at the surface of the lung) is generated when the inspired gas is 20% O_2, and is reduced in hypoxia.[33,34]

Three SODs have been cloned and characterized: cytosolic CuZnSOD, mitochondrial Mn-SOD, and extracellular SOD (EC-SOD). Cytosolic and mitochondrial forms of SOD constitute the majority of SOD activity in most tissues. EC-SOD is highly expressed in blood vessels, particularly arterial walls.[35–37] Vascular smooth muscle cells have been shown to generate large amounts of EC-SOD and it is thought that these cells are the principal source of the enzyme in the vascular wall.[37] The major portion of EC-SOD in the vasculature exists in the extracellular matrix. EC-SOD knockout mice develop normally, but when stressed by exposure to 99% O_2, these mutant mice survive less long than wild-type mice and have an earlier onset of severe lung oedema.[38] The role of SOD in signaling of oxygen tension will differ depending on the relative importance of $O_2^{\cdot-}$ and H_2O_2.

10.2.3. Mitochondrial Respiratory Chain

Mitochondria were proposed over 30 years ago as potential O_2 sensors in the carotid body.[39] Oxidative phosphorylation in mitochondria is carried out by four electrontransporting complexes (I–IV) and one H^+-translocating ATP synthetic complex (complex V). Mitochondrial generation of superoxide represents a byproduct of electron flow through the respiratory chain. Within the mitochondria, ROS production occurs primarily at two sites: complex I (NADH dehydrogenase) and complex III (ubiquinone-cytochrome bc1). The actions of inhibitors at specific sites in the electron transport chain have demonstrated that these sites are potentially important in the generation of ROS that initiate the redox signals that control the response to changes in oxygen tension. There are several studies providing further support for involvement of mitochondria in O_2-sensing in Hep 3B cells,[40–42] cardiomyocytes,[42–44] and in human ductus arteriosus.[45]

Complementary to these findings, Waypa et al. have recently shown that proximal but not distal inhibitors, acting in the electron transport chain (ETC) of mitochondria, suppressed HPV without affecting the response to other vasoconstrictors. They proposed that mitochondrial complex III is important in HPV, increasing ROS production under hypoxia and that generation of H_2O_2 is a potential mechanism by which pulmonary arterial tone could be regulated by the oxygen availability.[5,46] On the other hand, ETC inhibitors have been reported to mimic hypoxia by reducing ROS and causing pulmonary vasoconstriction.[17,47] Similarly, in the DA and renal arteries the inhibitors rotenone and antimycin A also mimic hypoxia by causing an increase in K^+ current and dilatation.[16,45] While it is agreed that mitochondrial ROS may help to mediate the vascular responses to hypoxia, whether hypoxia increases or decreases ROS remains to be solved.

10.3. Redox Sensitive Targets in Signaling Cascade

10.3.1. Ion Homeostasis

10.3.1.1. Ion Channels

Redox modulation of ion channel activity seems to be an important regulatory mechanism under physiological conditions for vasomotor functions. Several studies have provided evidence that O_2-sensing in vessels is mediated by effector potassium (K^+) channels with cytosolic redox as a sensor. The membrane potential of arterial smooth muscle cells is an important regulator of arterial tone and hence arterial diameter. These cells have steady or slowly changing resting membrane potential around −65 to −50 mV in vitro, close to the predicted equilibrium potential for K^+ ions (with physiological extracellular K^+ (~5 mM) the E_k is approximately −85 mV).[48] The opening of K^+ channels in the cell membrane increases K^+ efflux, which causes membrane hyperpolarization. This closes voltage-dependent Ca^{2+}-channels, decreasing Ca^{2+} entry and leading to vasodilatation. Conversely, inhibition of K^+ channels activity causes membrane depolarization, leads to Ca^{2+} entry, cell contraction and vasoconstriction.

The large conductance Ca^{2+}-activated potassium channels in PASMC and smooth muscle cells from systemic arteries, such as ear ASMC (EASMC), have been reported to be hypoxia and redox sensitive.[49] Reducing agents (DDT, GSH, and NADH) decreased PASMC, but not EASMC, K_{Ca} channel activity. However, oxidizing agents (DTNB, GSSG, and NAD) increased K_{Ca} channel activity in both PASMC and EASMC. The increased activity due to oxidizing agents was diminished by applying reducing agents. From these results, the authors suggest that the basal redox state of the PASMC K_{Ca} channel is more oxidized than that of the EASMC channel, since the response of K_{Ca} channels of the PASMC to intracellular reducing agents differs from that of the EASMC. In contrast, in isolated SMC from large conduit pulmonary arteries the K_{Ca} channel activity was unaffected by NAD and GSSG, or NADH and GSH.[50] This suggests that the change in the intracellular redox state, which occurs during acute hypoxia, does not alter the activity of K_{Ca} channels in SMC from large conduit pulmonary arteries. The different behavior of these channels in response to redox changes could be explained by the heterogeneity of the K_{Ca} in different tissues and may be related to the different responses of PASMC and EASMC K_{Ca} channels to hypoxia.

In SMC from resistance PA, the K^+ channels which control resting membrane potential (E_m), and inhibition of which initiates hypoxic pulmonary vasoconstriction, conduct a voltage-sensitive, delayed rectifying, outward current (Kv). This current activates at potentials less negative than −50 mV, is slowly inactivating and blocked by the Kv channel blocker 4-aminopyridine (4-AP) but not by inhibitors of K_{Ca} or K_{ATP} channels.[51–53] Park et al. showed, that DTT partially blocks Kv current in PASMC and accelerates the inactivation kinetics, but does not affect steady-state activation and inactivation[54]. Conversely, the oxidizing agent DTNBP increases Kv current and accelerates activation kinetics.[54]

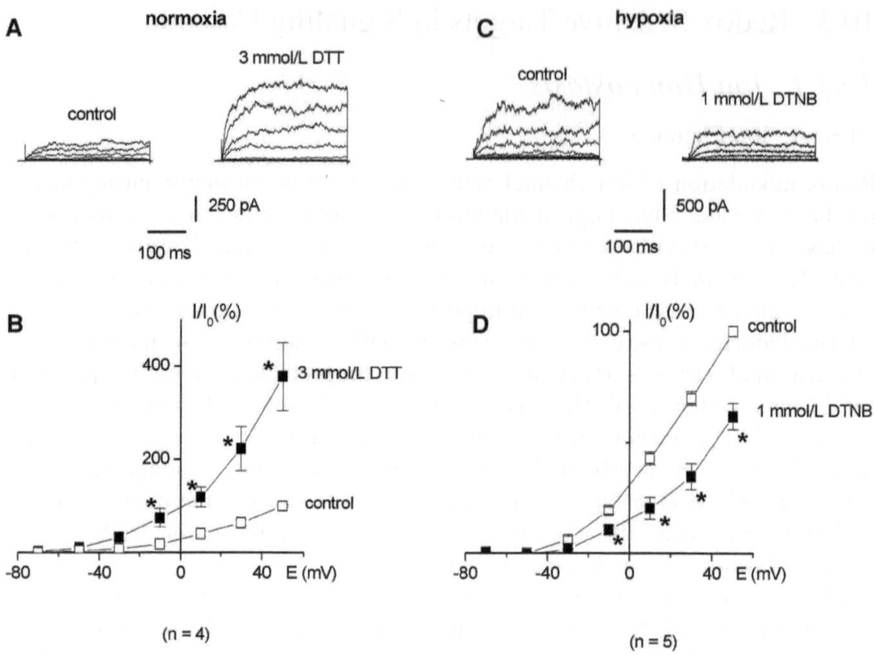

FIGURE 10.1. DTT activates and DTNB inhibits the whole-cell outward potassium current of SMC from fetal rabbit ductus arteriosus (DASMC). **(A)** Representative traces demonstrate potassium currents under normoxic conditions (left) and after 8-min exposure to 3 mmol/L DTT (right). **(B)** Averaged whole-cell I-V relationship of outward potassium currents normalized to maximum current at +50mV under normoxia (I/I_0) and after 8-min exposure to 3 mmol/L DTT. **(C)** Representative traces show potassium currents under hypoxic conditions (left) and after 8-min exposure to 1 mmol/L DTNB (right). **(D)** Averaged whole-cell I-V relationship of outward potassium currents normalized to maximum current at +50mV under hypoxia (I/I_0) and after 8-min exposure to 1 mmol/L DTNB. Currents were evoked from a holding potential of −70 mV to +50 mV in 20 mV steps. Values are mean ± SEM. Numbers of cells shown in parentheses. (*$p < 0.05$ for difference from control.) (From Olschewski et al.,[56] with permission.)

In initial experiments we reported that oxidized glutathione (GSSG), and diamide increased whole-cell K^+ current and hyperpolarized the resting membrane potential, while duroquinone did the opposite.[6,55] More recently, we have provided further support for the role of a redox-based O_2-sensing system in SMC from resistance pulmonary vessels and from the ductus arteriosus (DA), showing that the same redox signal elicits opposite effects on whole cell potassium current, membrane potential, cytosolic calcium and vascular tone in the resistance fetal pulmonary arteries and in DA.[56] (Figs 10.1–10.5). The ability of these redox agents to elicit exactly opposite responses in the DA and PA is consistent with a role for redox changes in their opposite vascular responses to any changes in O_2 tension.

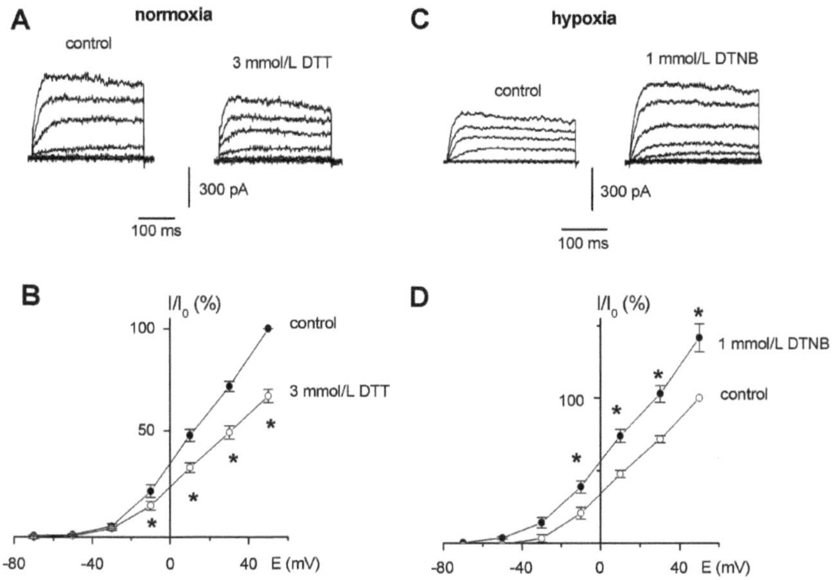

FIGURE 10.2. Effect of DTT and DTNB on whole-cell outward potassium current of fetal rabbit PASMC. **(A)** Representative traces demonstrate potassium currents under normoxic conditions (left) and after 8-min exposure to 3 mmol/L DTT (right). **(B)** Averaged whole-cell I–V relationship of outward potassium currents normalized to maximum current at +50mV under normoxia (I/I_0) and after 8-min exposure to 3 mmol/L DTT. **(C)** Representative traces show potassium currents under hypoxic conditions (left) and after 8-min exposure to 1 mmol/L DTNB (right). **(D)** Averaged whole-cell I–V relationship of outward potassium currents normalized to maximum current at +50mV under hypoxia (I/I_0) and after 8-min exposure to 1 mmol/L DTNB. Currents were activated by 20-mV voltage steps from a holding potential of −70 mV to +50 mV. Values are mean ± SEM. Numbers of cells shown in parentheses. (*$p < 0.05$ for difference from control.) (From Olschewski et al.,[56] with permission.)

Although the actual links between ROS and the function of ion channels are generally not well understood, the studies presented here suggest that vascular tone can be modulated by electron-transfer at the level of K^+ channels and that the redox status of thiols that control the activities of the ion channels may be a mechanism of regulation. Thus, vascular ion channels are controlled by multiple redox-linked mechanisms and this is likely to be responsible for the diversity of observations that have been made.

10.3.1.2. Modulation of Cytosolic Ca^{2+} Concentration

Ca^{2+} is a second messenger that regulates a variety of biological processes, including gene expression, neurotransmission, cell motility, and cell growth. In response to physiological stimuli at the cell surface, the intracellular level of Ca^{2+} may rise, and this elevation elicit the activation of Ca^{2+}-dependent proteins such as

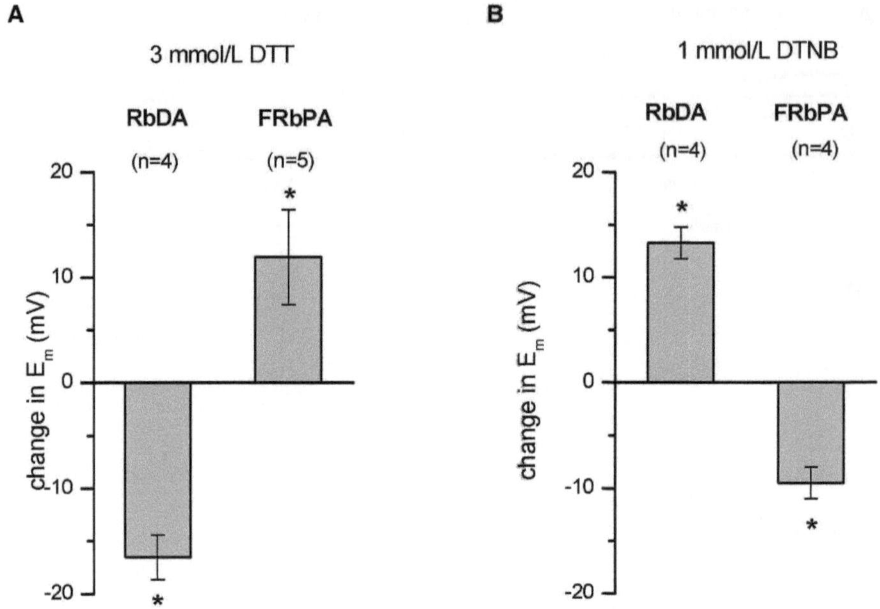

FIGURE 10.3. DTT and DTNB modulation of resting membrane potential (E_m) under normoxic (**A**) and hypoxic conditions (**B**). Values are mean ± SEM of the change in E_m measured with current-clamp ($I = 0$) in rabbit DA smooth muscle cells (RbDA), and fetal rabbit PA smooth muscle cells (FrbPA). Numbers of cells shown in parentheses. (*$p < 0.05$ for difference from control.) (From Olschewski et al.,[56] with permission.)

protein kinase C, Ca^{2+}-calmodulin kinases, and calmodulin-dependent protein phosphatases. Oxidants have been shown to stimulate Ca^{2+} signaling by increasing cytosolic Ca^{2+} concentration in cardiac myocytes,[57] suggesting a possible physiological role for ROS and oxidant stress in the regulation of Ca^{2+}-induced signaling.

Abundant evidence indicates that oxidant stress in the endothelium increases the intracellular Ca^{2+}concentration ($[Ca^{2+}]_i$) which in turn correlates with increased endothelial permeability. Direct treatment of venous endothelial cells with H_2O_2 increased the $[Ca^{2+}]_i$ showing a transient release of Ca^{2+} from the inositol triphosphate(IP_3)-sensitive intracellular stores (Doan, 1994). In addition, H_2O_2 was found to regulate intracellular Ca^{2+} signaling by stimulating Ca^{2+} release from the IP_3-sensitive stores in bovine pulmonary endothelial cells,[58] and human endothelial cells.[59] Similarly, in rabbit aortic endothelial cells, treatment with linoleic acid hydroperoxide resulted in a transient increase in intracellular Ca^{2+}.[60] The increased influx mediated by xanthine/xanthine oxidase in aortic endothelial cells has been reported to be inhibited by an anion channel blocker, Ni^{2+} (an inorganic membrane Ca^{2+} channel blocker), dithiothreitol (a reducing agent), and inhibitors of the Haber-Weiss reaction.[61] These observations suggest that Ca^{2+} influx occurred through membrane Ca^{2+} channels that might be regulated by $OH^{\bullet}$

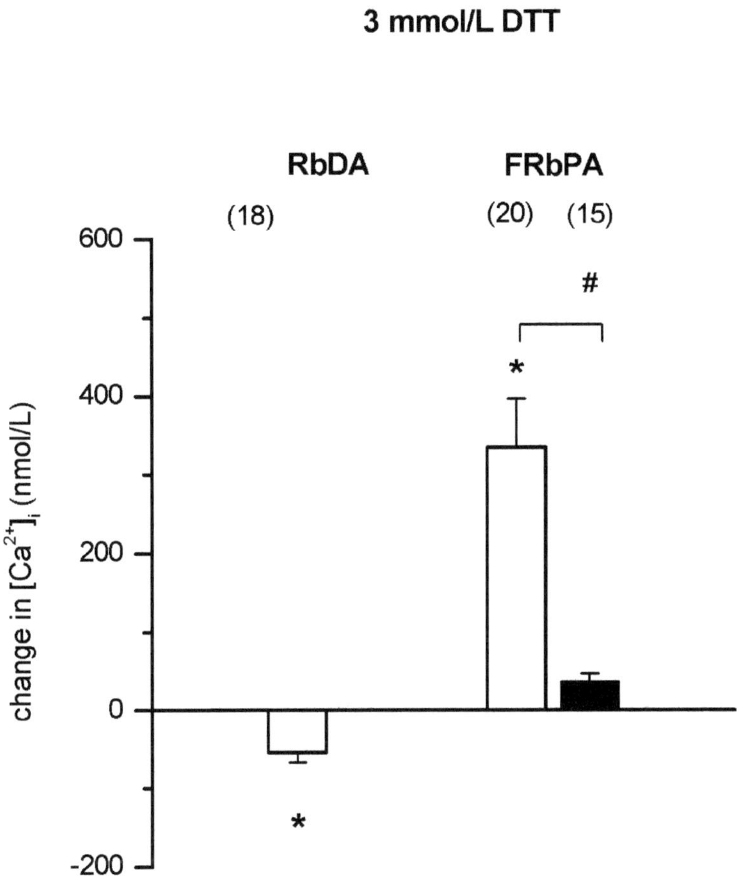

FIGURE 10.4. Changes in intracellular Ca^{2+} concentration of DA smooth muscle cells (RbDA) and fetal rabbit PA smooth muscle cells (FrbPA) after applications of 3 mmol/L DTT (white bars). The black bar indicates the changes in intracellular Ca^{2+} concentration of fetal rabbit PA smooth muscle cells (FRbPA) after pretreatment of cells with 10 μmol/L nifedipine and 3 mmol/L DTT. Number of cells in parenthesis. Values are mean ± SEM. ($^*p < 0.05$ compared with baseline. $^\#p < 0.05$ compared with DTT response alone.) (From Olschewski et al.,[56] with permission.)

generation. Shasby et al. reported that the Ca^{2+} entry blocker $LaCl_3$ inhibited the increase in albumin permeability of porcine pulmonary artery endothelial cells, seen in response to ROS generated by xanthine/xanthine oxidase,[62] yet it was ineffective in inhibition of the H_2O_2-mediated increased permeability of bovine pulmonary microvascular endothelial cells.[63] These contrasting findings may be attributable to the species difference of the cells, different levels or types of ROS or may be related to different regulatory mechanisms of permeability utilized by endothelial cells from conduit vs. microvascular vessels.

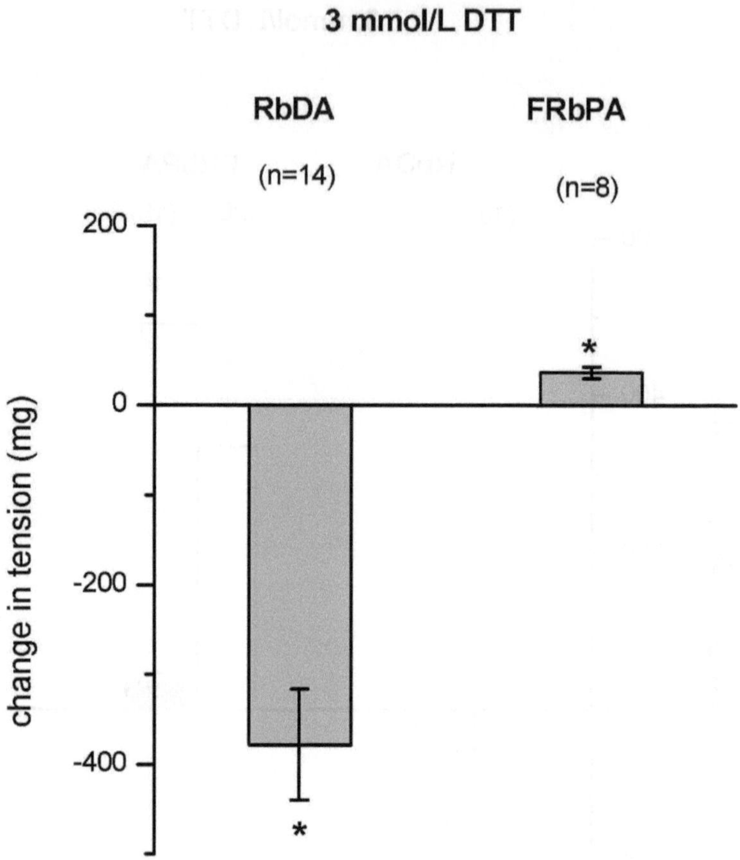

FIGURE 10.5. DTT modulation of tone in vascular rings. Change in vascular tone of DA (RbDA) and fetal rabbit PA (FRbPA) rings after application of 3 mmol/L DTT under normoxic conditions. Numbers of measurements shown in parentheses. Values are mean ± SEM. (*$p < 0.05$ for difference from control.) (From Olschewski et al.,[56] with permission.)

In skeletal muscle, Feng et al. have provided direct evidence that SR-ryanodine channel activity follows the transmembrane redox potential, by co-localization of the transporter selective for glutathione with the ryanodine receptor (RyR).[64] The molecular basis of O_2-responsiveness in these cells was identified to involve 6-8 of 50 RyR thiols whose redox state is dynamically controlled, as described by Eu et al.[65]

Biochemical evidence suggests that the Ca^{2+}-release channel of cardiac SR can also be regulated by the redox potential. Boraso and Williams demonstrated that H_2O_2 increased the open probability of the Ca^{2+}-release channel in sheep and the channel activity was suppressed by DTT[66]. The effect of H_2O_2 and DTT on intracellular [Ca^{2+}] in cardiomyocytes is consistent with our observations on the tension of the ductus arteriosus (DA).[8] H_2O_2 constricted and DTT dilated

DA rings. The opposite effects of H_2O_2 and DTT on DA rings would indicate a potential redox mechanism for normoxia-induced vasoconstriction and hypoxia-induced vasodilatation in DA.

One of the mechanisms by which redox changes are able to regulate Ca^{2+} release from the SR is via thiol-disulfide exchange reactions, converting the disulfide structure to another thiol through reduction (by GSH or DTT) reactions. This is consistent with the proposal by Abramson and Salama in skeletal muscle that thiol-disulfide interchange reactions within the Ca^{2+}-release channel molecule convert the open/closed states of the channel[67,68] or that thiol reductants DTT and GSH suppress SR Ca^{2+}-release channel activity[66,69]. Another redox-sensitive mechanism is presented by Wilson et al. in PASMC, involving Ca^{2+}-release from ryanodine-sensitive SR stores. Acute hypoxia increases β-NADH levels (i.e., more reduced), which then increase the net amount of cyclic ADP-ribose (cADPR), and simultaneously inhibit cADP degradation. The increased level of cADPR promotes Ca^{2+}-release from RyR and elicits vasoconstriction.[70]

Although the exact molecular targets of redox-mediated Ca^{2+} signaling are not known, there are at least three mechanisms by which redox could alter Ca^{2+} flux.

1. *Ca^{2+} pumps:* The ability of various oxidants to inhibit the activity of an ATP-dependent Ca^{2+} pump[71,72] suggests that direct modification of Ca^{2+} pumps by oxidants may be one mechanism of oxidant-mediated Ca^{2+} signaling.
2. *Ca^{2+}channel:* It is also likely that sensitive thiol-groups within the complexes are an essential component of a transmembrane redox sensor and may be involved in mediating changes in Ca^{2+} signaling during changes in oxygen tension. In this case, oxidation produces a significant increase in Ca^{2+}-release.
3. *Release:* Reversible activation of cardiac RyR channels by oxidation could be relevant in the heart, especially when there is an increase in free radical production such as in ischemia/reperfusion situations. Thus, evidence exists to suggest the potential involvement of ROS-mediated Ca^{2+} flux in early signaling pathways; however, the underlying cellular and molecular mechanisms remain to be solved.

10.4. Reactive Oxygen Species: Implications for Physiological Responses

10.4.1. Hypoxic Pulmonary Vasoconstriction

The pulmonary vascular bed is unique compared with most systemic vascular beds. One of the characteristics that distinguishes the pulmonary circulation from the systemic circulation is its sensitivity to oxygen. The pulmonary circulation responds to low oxygen tensions by constricting—hypoxic pulmonary vasoconstriction (HPV).[73,74] This HPV diverts mixed venous blood away from hypoxic alveoli, thus optimizing the matching of perfusion and ventilation and preventing arterial hypoxemia. In contrast, the systemic circulation dilates in response to hypoxia and thus, increases blood flow to under-oxygenated regions. In the healthy lung, when only a small region of the lung is hypoxic (e.g., in atelectasis), HPV can occur without significant effect on pulmonary arterial pressure. However, when hypoxia is generalized, as seen with many lung diseases and in high-altitude exposure, the subsequent pulmonary vasoconstriction contributes to pulmonary hypertension. Under such conditions, HPV may have profound hemodynamic consequences, including a reduction in cardiac output and right ventricular failure.

O_2 acts as a vasodilator of the pulmonary circulation and O_2 therapy has a modest effect in reducing pulmonary hypertension in hypoxic lung disease. H_2O_2 is one of the most stable reactive oxygen products, produced by the rapid dismutation of superoxide radicals. Because H_2O_2 alters pulmonary vasoreactivity.[2,75,76] generation of H_2O_2 is a potential mechanism by which pulmonary arterial tone can be regulated by oxygen availability. Alterations in K^+ channel activity play an important role in the development of hypoxic pulmonary vasoconstriction.[6] Under normoxia, basal production of H_2O_2 results in an oxidized cellular redox state and outward K^+ current is maintained. However, the decrease in oxygen partial pressure from normoxic to hypoxic levels also decreases the pulmonary production of H_2O_2 and shifts the ratio of redox couples toward a more reduced state in PASMC, leading to closure of Kv channels, membrane depolarization, L-type Ca^{2+} channel activation, an increase in cytosolic $[Ca^{2+}]$ and vasoconstriction. Modulating this cascade, the large conductance K_{Ca} channels are activated by intracellular Ca^{2+} and membrane depolarization. The opening of these K_{Ca} channels tends to hyperpolarize the cell membrane, opposing the effect of Kv channels closure and thus preventing excessive vasoconstriction.

10.4.2. Transcription Factors Affected by ROS

Eukaryotic cells have developed highly elaborate mechanisms to rapidly respond to changes in their environment by altering the expression of genes. One form of cellular stress is the production of ROS. The notion that oxidant stress can modulate a wide variety of biological processes by coupling signals at the cell surface into long-term changes in gene expression suggests that multiple signaling pathways are involved. ROS regulate several general classes of genes, including those coding adhesion molecules and chemotactic factors, antioxidant enzymes and vasoactive substances. Transcription factors are the principal nuclear factors that control gene expression. ROS may be defined as second-messenger molecules that regulate various intracellular signal transduction cascades and ultimately affect transcriptional activity. Indeed, ROS can affect multiple signal transduction pathways upstream of nuclear transcription factors, including modulation of Ca^{2+} signaling, protein kinase, and protein phosphatase pathways.

Redox-sensitive modulation of transcription factor activity can occur via (1) direct oxidative modification of the transcription factor itself by intracellular ROS or (2) posttranslational modifications (i.e., phosphorylation/dephosphorylation), by the effects of redox-regulated intracellular signaling cascades. One of the best characterized transcription factors responsive to redox regulation is NF-κB. *In vitro* studies have already documented the involvement of this transcription factor in tubular morphogenesis of human microvascular endothelial cells induced by oxidative stress.[77] NF-κB may be involved in the pathogenesis of proliferative disorders of the vasculature, including restenosis and arteriosclerosis.[78,79] Several studies have demonstrated the presence of NF-κB binding sites in the promoters of human and mouse VEGF genes. Expression of a potent angiogenic gene, VEGF, was found to be upregulated by hypoxia of newborn rat heart cells.[80] H_2O_2-mediated strong VEGF gene expression has been demonstrated in endothelial cells,[81] indicating that ROS signaling can potentially alter changes in gene expression.

Acknowledgments. E. Kenneth Weir is supported by VA Merit Review Funding and NIH (RO1-HL 65322-01A1). Andrea Olschewski is supported by the Deutsche Forschungsgemeinschaft (DFG: SFB 547). The authors thank Brigitte Agari for excellent technical assistance.

References

1. Archer, S.L., Will, J.A., and Weir, E.K., Redox status in the control of pulmonary vascular tone, *Herz* **11**(3), 127–141 (1986).
2. Burke, T.M., and Wolin, M.S., Hydrogen peroxide elicits pulmonary arterial relaxation and guanylate cyclase activation, *Am. J. Physiol.* **252**(4), H721–H732 (1987).
3. Wolin, M.S., Burke-Wolin, T.M., and Mohazzab, H., Roles for NAD(P)H oxidases and reactive oxygen species in vascular oxygen sensing mechanisms, *Respir. Physiol.* **115**(2), 229–238 (1999).
4. Archer, S.L., Weir, E.K., Reeve, H.L., and Michelakis, E., Molecular identification of O_2 sensors and O_2-sensitive potassium channels in the pulmonary circulation, *Adv. Exp. Med. Biol.* **475**, 219–240 (2000).
5. Waypa, G.B., Chandel, N.S., and Schumacker, P.T., Model for hypoxic pulmonary vasoconstriction involving mitochondrial oxygen sensing, *Circ. Res.* **88**(12), 1259–1266 (2001).
6. Weir, E.K., Archer, S.L., The mechanism of acute hypoxic pulmonary vasoconstriction: the tale of two channels, *FASEB J.* **9**(2), 183–189 (1995).
7. Post, J.M., Hume, J.R., Archer, S.L., and Weir, E.K., Direct role for potassium channel inhibition in hypoxic pulmonary vasoconstriction, *Am. J. Physiol.* **262**(4 Pt 1), C882–C890 (1992).
8. Reeve, H.L., Tolarova, S., Nelson, D.P., Archer, S., and Weir, E.K., Redox control of oxygen sensing in the rabbit ductus arteriosus, *J. Physiol.* **533**(Pt 1), 253–261 (2001).
9. Yuan, X.J., Goldman, W.F., Tod, M.L., Rubin, L.J., and Blaustein, M.P., Hypoxia reduces potassium currents in cultured rat pulmonary but not mesenteric arterial myocytes, *Am. J. Physiol.* **264**(2 Pt 1), L116–L123 (1993).
10. McMurtry, I.F., Davidson, A.B., Reeves, J.T., and Grover, R.F., Inhibition of hypoxic pulmonary vasoconstriction by calcium antagonists in isolated rat lungs, *Circ. Res.* **38**(2), 99–104 (1976).
11. McMurtry, I.F., BAY K 8644 potentiates and A23187 inhibits hypoxic vasoconstriction in rat lungs, *Am. J. Physiol.* **249**(4 Pt 2), H741–H746 (1985).

12. Tolins, M., Weir, E.K., Chesler, E., Nelson, D.P., and From, A.H., Pulmonary vascular tone is increased by a voltage-dependent calcium channel potentiator, *J. Appl. Physiol.* **60**(3), 942–948 (1986).
13. Nakanishi, T., Gu, H., Hagiwara, N., and Momma, K., Mechanisms of oxygen-induced contraction of ductus arteriosus isolated from the fetal rabbit, *Circ. Res.* **72**(6), 1218–1228 (1993).
14. Roulet, M.J., Coburn, R.F., Oxygen-induced contraction in the guinea pig neonatal ductus arteriosus, *Circ. Res.* **49**(4), 997–1002 (1981).
15. Tristani-Firouzi, M., Reeve, H.L., Tolarova, S., Weir, E.K., and Archer, S.L., Oxygen-induced constriction of rabbit ductus arteriosus occurs via inhibition of a 4-aminopyridine-, voltage-sensitive potassium channel, *J. Clin. Invest.* **98**(9), 1959–1965 (1996).
16. Michelakis, E.D., Hampl, V., Nsair, A., Wu, X., Harry, G., Haromy, A., Gurtu, R., and Archer, S.L., Diversity in mitochondrial function explains differences in vascular oxygen sensing, *Circ. Res.* **90**(12), 1307–1315 (2002).
17. Archer, S.L., Wu, X.C., Thebaud, B., Moudgil, R., Hashimoto, K., and Michelakis, E.D., O_2 sensing in the human ductus arteriosus: redox- sensitive K^+ channels are regulated by mitochondria- derived hydrogen peroxide, *Biol. Chem.* **385**(3–4), 205–216 (2004).
18. Youngson, C., Nurse, C., Yeger, H., and Cutz, E., Oxygen sensing in airway chemoreceptors, *Nature* **365**(6442), 153–155 (1993).
19. Zulueta, J.J., Yu, F.S., Hertig, I.A., Thannickal, V.J., and Hassoun, P.M., Release of hydrogen peroxide in response to hypoxia-reoxygenation: role of an NAD(P)H oxidase-like enzyme in endothelial cell plasma membrane, *Am. J. Respir. Cell Mol. Biol.* **12**(1), 41–49 (1995).
20. Kummer, W., Acker, H., Immunohistochemical demonstration of four subunits of neutrophil NAD(P)H oxidase in type I cells of carotid body, *J. Appl. Physiol.* **78**(5), 1904–1909 (1995).
21. Mohazzab, K.M., Wolin, M.S., Properties of a superoxide anion-generating microsomal NADH oxidoreductase, a potential pulmonary artery PO_2 sensor, *Am. J. Physiol.* **267**(6 Pt 1), L823–L831 (1994).
22. Griendling, K.K., Minieri, C.A., Ollerenshaw, J.D., and Alexander, R.W., Angiotensin II stimulates NADH and NADPH oxidase activity in cultured vascular smooth muscle cells, *Circ. Res.* **74**(6), 1141–1148 (1994).
23. Pagano, P.J., Tornheim, K., and Cohen, R.A., Superoxide anion production by rabbit thoracic aorta: effect of endothelium-derived nitric oxide, *Am. J. Physiol.* **265**(2), H707–H712 (1993).
24. Pagano, P.J., Ito, Y., Tornheim, K., Gallop, P.M., Tauber, A.I., and Cohen, R.A., An NADPH oxidase superoxide-generating system in the rabbit aorta, *Am. J. Physiol.* **268**(6), H2274–H2280 (1995).
25. Griendling, K.K., Ushio-Fukai, M., Redox control of vascular smooth muscle proliferation, *J. Lab. Clin. Med.* **132**(1), 9–15 (1998).
26. Ohara, Y., Peterson, T.E., and Harrison, D.G., Hypercholesterolemia increases endothelial superoxide anion production, *J. Clin. Invest.* **91**(6), 2546–2551 (1993).
27. Pagano, P.J., Chanock, S.J., Siwik, D.A., Colucci, W.S., and Clark, J.K., Angiotensin II induces p67phox mRNA expression and NADPH oxidase superoxide generation in rabbit aortic adventitial fibroblasts, *Hypertension* **32**(2), 331–337 (1998).
28. Wang, D., Youngson, C., Wong, V., Yeger, H., Dinauer, M.C., Vega-Saenz, M.E., Rudy, B., and Cutz, E., NADPH-oxidase and a hydrogen peroxide-sensitive K^+ channel may

function as an oxygen sensor complex in airway chemoreceptors and small cell lung carcinoma cell lines, *Proc. Natl. Acad. Sci. USA.* **93**(23), 13182–13187 (1996).

29. Archer, S.L., Reeve, H.L., Michelakis, E., Puttagunta, L., Waite, R., Nelson, D.P., Dinauer, M.C., and Weir, E.K., O_2 sensing is preserved in mice lacking the gp91 phox subunit of NADPH oxidase, *Proc. Natl. Acad. Sci. USA.* **96**, 7944–7949 (1999).
30. Marshall, C., Mamary, A.J., Verhoeven, A.J., and Marshall, B.E., Pulmonary artery NADPH-oxidase is activated in hypoxic pulmonary vasoconstriction, *Am. J. Respir. Cell Mol. Biol.* **15**(5), 633–644 (1996).
31. Weissmann, N., Tadic, A., Hanze, J., Rose, F., Winterhalder, S., Nollen, M., Schermuly, R.T., Ghofrani, H.A., Seeger, W., and Grimminger, F., Hypoxic vasoconstriction in intact lungs: a role for NADPH oxidase-derived H_2O_2?, *Am. J. Physiol. Lung Cell. Mol. Physiol.* **279**(4), L683–L690 (2000).
32. Nozik-Grayck, E., Piantadosi, C.A., van, A.J., Alper, S.L., and Huang, Y.C., Protection of perfused lung from oxidant injury by inhibitors of anion exchange, *Am. J. Physiol.* **273**(2), L296–L304 (1997).
33. Archer, S.L., Nelson, D.P., and Weir, E.K., Simultaneous measurement of O_2 radicals and pulmonary vascular reactivity in rat lung, *J. Appl. Physiol.* **67**(5), 1903–1911 (1989).
34. Paky, A., Michael, J.R., Burke-Wolin, T.M., Wolin, M.S., and Gurtner, G.H., Endogenous production of superoxide by rabbit lungs: effects of hypoxia or metabolic inhibitors, *J. Appl. Physiol.* **74**(6), 2868–2874 (1993).
35. Oury, T.D., Chang, L.Y., Marklund, S.L., Day, B.J., and Crapo, J.D., Immunocytochemical localization of extracellular superoxide dismutase in human lung, *Lab. Invest.* **70**(6), 889–898 (1994).
36. Oury, T.D., Day, B.J., and Crapo, J.D., Extracellular superoxide dismutase in vessels and airways of humans and baboons, *Free Radic. Biol. Med.* **20**(7), 957–965 (1996).
37. Stralin, P., Karlsson, K., Johansson, B.O., and Marklund, S.L., The interstitium of the human arterial wall contains very large amounts of extracellular superoxide dismutase, *Arterioscler. Thromb. Vasc. Biol.* **15**(11), 2032–2036 (1995).
38. Carlsson, L.M., Jonsson, J., Edlund, T., and Marklund, S.L., Mice lacking extracellular superoxide dismutase are more sensitive to hyperoxia, *Proc. Natl. Acad. Sci. USA.* **92**(14), 6264–6268 (1995).
39. Mills, E., Jobsis, F.F., Mitochondrial respiratory chain of carotid body and chemoreceptor response to changes in oxygen tension, *J. Neurophysiol.* **35**(4), 405–428 (1972).
40. Chandel, N.S., Maltepe, E., Goldwasser, E., Mathieu, C.E., Simon, M.C., and Schumacker, P.T., Mitochondrial reactive oxygen species trigger hypoxia-induced transcription, *Proc. Natl. Acad. Sci. USA.* **95**(20), 11715–11720 (1998).
41. Chandel, N.S., McClintock, D.S., Feliciano, C.E., Wood, T.M., Melendez, J.A., Rodriguez, A.M., and Schumacker, P.T., Reactive oxygen species generated at mitochondrial complex III stabilize hypoxia-inducible factor-1a during hypoxia: a mechanism of O_2 sensing, *J. Biol. Chem.* **275**(33), 25130–25138 (2000).
42. Chandel, N.S., Schumacker, P.T., Cellular oxygen sensing by mitochondria: old questions, new insight, *J. Appl. Physiol.* **88**(5), 1880–1889 (2000).
43. Budinger, G.R., Chandel, N.S., Shao, Z.H., Li, C.Q., Mehmed, A., Becker, L.B., and Schumacker, P.T., Cellular energy utilization and supply during hypoxia in embryonic cardiac myocytes, *Am. J. Physiol.* **270**, L44–L53 (1996).
44. Budinger, G.R., Duranteau, J., Chandel, N.S., and Schumacker, P.T., Hibernation during hypoxia in cardiomyocytes. Role of mitochondria as the O_2 sensor, *J. Biol. Chem.* **273**(6), 3330–3336 (1998).

45. Michelakis, E.D., Rebeyka, I., Wu, X.C., Nsair, A., Thébaud, B., Hashimoto, K., Dyck, J.R.B., Haromy, A., Harry, G., Barr, A., and Archer, S.L., O_2 sensing in the human ductus arteriosus—Regulation of voltage-gated K^+ channels in smooth muscle cells by a mitochondrial redox sensor, *Circ. Res.* **91**(6), 478–486 (2002).
46. Waypa, G.B., Marks, J.D., Mack, M.M., Boriboun, C., Mungai, P.T., and Schumacker, P.T., Mitochondrial reactive oxygen species trigger calcium increases during hypoxia in pulmonary arterial myocytes, *Circ. Res.* **91**(8), 719–726 (2002).
47. Archer, S.L., Huang, J., Henry, T., Peterson, D., and Weir, E.K., A redox-based O_2 sensor in rat pulmonary vasculature, *Circ. Res.* **73**(6), 1100–1112 (1993).
48. Nelson, M.T., Quayle, J.M., Physiological roles and properties of potassium channels in arterial smooth muscle, *Am. J. Physiol.* **268**(4 Pt 1), C799–C822 (1995).
49. Park, M.K., Lee, S.H., Lee, S.J., Ho, W.K., and Earm, Y.E., Different modulation of Ca-activated K channels by the intracellular redox potential in pulmonary and ear arterial smooth muscle cells of the rabbit, *Pflügers Arch.* **430**(3), 308–314 (1995).
50. Thuringer, D., Findlay, I., Contrasting effects of intracellular redox couples on the regulation of maxi-K channels in isolated myocytes from rabbit pulmonary artery, *J. Physiol.* **500**(Pt 3), 583–592 (1997).
51. Yuan, X.J., Voltage-gated K^+ currents regulate resting membrane potential and $[Ca^{2+}]_i$ in pulmonary arterial myocytes, *Circ. Res.* **77**(2), 370–378 (1995).
52. Archer, S.L., Huang, J.M., Reeve, H.L., Hampl, V., Tolarova, S., Michelakis, E., and Weir, E.K., Differential distribution of electrophysiologically distinct myocytes in conduit and resistance arteries determines their response to nitric oxide and hypoxia, *Circ. Res.* **78**(3), 431–442 (1996).
53. Archer, S.L., Souil, E., Dinh-Xuan, A.T., Schremmer, B., Mercier, J.C., Yaagoubi, A. El, Nguyen-Huu, L., Reeve, H.L., and Hampl, V., Molecular identification of the role of voltage-gated K^+ channels, Kv1.5 and Kv2.1, in hypoxic pulmonary vasoconstriction and control of resting membrane potential in rat pulmonary artery myocytes, *J. Clin. Invest.* **101**(11), 2319–2330 (1998).
54. Park, M.K., Bae, Y.M., Lee, S.H., Ho, W.K., and Earm, Y.E., Modulation of voltage-dependent K^+ channel by redox potential in pulmonary and ear arterial smooth muscle cells of the rabbit, *Pflügers Arch.* **434**(6), 764–771 (1997).
55. Reeve, H.L., Weir, E.K., Nelson, D.P., Peterson, D.A., and Archer, S.L., Opposing effects of oxidants and antioxidants on K^+ channel activity and tone in rat vascular tissue, *Exp. Physiol.* **80**(5), 825–834 (1995).
56. Olschewski, A., Hong, Z., Peterson, D.A., Nelson, D.P., Porter, V.A., and Weir, E.K., Opposite effects of redox status on membrane potential, cytosolic calcium, and tone in pulmonary arteries and ductus arteriosus, *Am. J. Physiol. Lung Cell. Mol. Physiol.* **286**(1), L15–L22 (2004).
57. Suzuki, Y.J., Cleemann, L., Abernethy, D.R., and Morad, M., Glutathione is a cofactor for H_2O_2-mediated stimulation of Ca^{2+}-induced Ca^{2+} release in cardiac myocytes, *Free Radic. Biol. Med.* **24**(2), 318–325 (1998).
58. Boyer, C.S., Bannenberg, G.L., Neve, E.P., Ryrfeldt, A., and Moldeus, P., Evidence for the activation of the signal-responsive phospholipase A2 by exogenous hydrogen peroxide, *Biochem. Pharmacol.* **50**(6), 753–761 (1995).
59. Dreher, D., Junod, A.F., Differential effects of superoxide, hydrogen peroxide, and hydroxyl radical on intracellular calcium in human endothelial cells, *J. Cell Physiol.* **162**(1), 147–153 (1995).

60. Sweetman, L.L., Zhang, N.Y., Peterson, H., Gopalakrishna, R., and Sevanian, A., Effect of linoleic acid hydroperoxide on endothelial cell calcium homeostasis and phospholipid hydrolysis, *Arch. Biochem. Biophys.* %**20;323**(1), 97–107 (1995).
61. Az-ma, T., Saeki, N., and Yuge, O., Cytosolic Ca^{2+} movements of endothelial cells exposed to reactive oxygen intermediates: role of hydroxyl radical-mediated redox alteration of cell-membrane Ca^{2+} channels, *Br. J. Pharmacol.* **126**(6), 1462–1470 (1999).
62. Shasby, D.M., Lind, S.E., Shasby, S.S., Goldsmith, J.C., and Hunninghake, G.W., Reversible oxidant-induced increases in albumin transfer across cultured endothelium: alterations in cell shape and calcium homeostasis, *Blood* **65**(3), 605–614 (1985).
63. Siflinger-Birnboim, A., Lum, H., Del Vecchio, P.J., and Malik, A.B., Involvement of Ca^{2+} in the H_2O_2-induced increase in endothelial permeability, *Am. J. Physiol.* **270**(6), L973–L978 (1996).
64. Feng, W., Liu, G., Allen, P.D., and Pessah, I.N., Transmembrane redox sensor of ryanodine receptor complex, *J. Biol. Chem.* **275**(46), 35902–35907 (2000).
65. Eu, J.P., Sun, J., Xu, L., Stamler, J.S., and Meissner, G., The skeletal muscle calcium release channel: coupled O_2 sensor and NO signaling functions, *Cell* **102**(4), 499–509 (2000).
66. Boraso, A., Williams, A.J., Modification of the gating of the cardiac sarcoplasmic reticulum Ca^{2+}-release channel by H_2O_2 and dithiothreitol, *Am. J. Physiol.* **267**(3 Pt 2), H1010–H1016 (1994).
67. Abramson, J.J., Salama, G., Sulfhydryl oxidation and Ca^{2+} release from sarcoplasmic reticulum, *Mol. Cell. Biochem.* **82**(1–2), 81–84 (1988).
68. Abramson, J.J., Salama, G., Critical sulfhydryls regulate calcium release from sarcoplasmic reticulum, *J. Bioenerg. Biomembr.* **21**(2), 283–294 (1989).
69. Zaidi, N.F., Lagenaur, C.F., Abramson, J.J., Pessah, I., and Salama, G., Reactive disulfides trigger Ca^{2+} release from sarcoplasmic reticulum via an oxidation reaction, *J. Biol. Chem.* **264**(36), 21725–21736 (1989).
70. Wilson, H.L., Dipp, M., Thomas, J.M., Lad, C., Galione, A., and Evans, A.M., Adp-ribosyl cyclase and cyclic ADP-ribose hydrolase act as a redox sensor. a primary role for cyclic ADP-ribose in hypoxic pulmonary vasoconstriction, *J. Biol. Chem.* **276**(14), 11180–11188 (2001).
71. Grover, A.K., Samson, S.E., Effect of superoxide radical on Ca^{2+} pumps of coronary artery, *Am. J. Physiol.* **255**(3), C297–C303 (1988).
72. Grover, A.K., Samson, S.E., and Fomin, V.P., Peroxide inactivates calcium pumps in pig coronary artery, *Am. J. Physiol.* **263**(2), H537–H543 (1992).
73. Bradford, J.R., Dean, H.P., The pulmonary circulation, *J. Physiol.* **16**, 34–96 (1894).
74. Euler v, U.S., Liljestrand, G., Observations on the pulmonary arterial blood pressure in the cat, *Acta Physiol. Scand.* **12**, 301–320 (1946).
75. Burke-Wolin, T., Abate, C.J., Wolin, M.S., and Gurtner, G.H., Hydrogen peroxide-induced pulmonary vasodilation: role of guanosine 3′,5′-cyclic monophosphate, *Am. J. Physiol.* **261**(6), L393–L398 (1991).
76. Archer, S.L., Hampl, V., Nelson, D.P., Sidney, E., Peterson, D.A., and Weir, E.K., Dithionite increases radical formation and decreases vasoconstriction in the lung. Evidence that dithionite does not mimic alveolar hypoxia, *Circ. Res.* **77**(1), 174–181 (1995).
77. Shono, T., Ono, M., Izumi, H., Jimi, S.I., Matsushima, K., Okamoto, T., Kohno, K., and Kuwano, M., Involvement of the transcription factor NF-kB in tubular morphogenesis

of human microvascular endothelial cells by oxidative stress, *Mol. Cell. Biol.* **16**(8), 4231–4239 (1996).
78. Brand, K., Page, S., Walli, A.K., Neumeier, D., and Baeuerle, P.A., Role of nuclear factor-kB in atherogenesis, *Exp. Physiol.* **82**(2), 297–304 (1997).
79. Maruyama, I., Shigeta, K., Miyahara, H., Nakajima, T., Shin, H., Ide, S., and Kitajima, I., Thrombin activates NF-k B through thrombin receptor and results in proliferation of vascular smooth muscle cells: role of thrombin in atherosclerosis and restenosis, *Ann. NY Acad. Sci.* **811**, 429–436 (1997).
80. Ladoux, A., Frelin, C., Hypoxia is a strong inducer of vascular endothelial growth factor mRNA expression in the heart, *Biochem. Biophys. Res. Commun.* **195**(2), 1005–1010 (1993).
81. Chua, C.C., Hamdy, R.C., and Chua, B.H., Upregulation of vascular endothelial growth factor by H_2O_2 in rat heart endothelial cells, *Free Radic. Biol. Med.* **25**(8), 891–897 (1998).

11
Impedance Spectroscopy and Quartz Crystal Microbalance

Noninvasive Tools to Analyze Ligand–Receptor Interactions at Functionalized Surfaces and of Cell Monolayers

Andreas Hinz and Hans-Joachim Galla*

11.1. Introduction

In recent decades, analytical tools to investigate biological samples have been improved to gain better detection limits and deeper insights into molecular interactions. Two of the main developments, the impedance spectroscopy and the quartz crystal microbalance technique, concern the characterization of surface activity of biomolecules such as phospholipids, proteins, or DNA/RNA without sample invasion. By means of impedance spectroscopy the electrical conductance of solid supported membranes can be measured, that is, ion transport across and protein adsorption onto membranes can be characterized. Furthermore, *in vivo* studies of cell adhesion, cell proliferation and cell-cell interactions are possible. By means of quartz crystal microbalance ligand-receptor and protein-membrane interactions can be analyzed. Both applications are related in their physical and mathematical background and are presented in the following overview with the above-mentioned applications.

11.2. Impedance Spectroscopy of Biomolecules

Impedance spectroscopy has become an important noninvasive tool to investigate lipid membranes as well as cell monolayers on a solid support.[1] We would like to introduce these two different kinds of systems, which show the powerful applications of this technique. The interaction of membrane-bound receptors and solubilized ligands can be characterized as a model of specific lipid-protein interactions, and the transport activity of membrane-anchored channels on positively or negatively charged ions can be analyzed.[2] The second relatively new approach is the investigation of cell monolayers onto a solid substrate. Electric parameters are

* Institut für Biochemie, Westfälische Wilhelms-Universität Münster, 48149 Münster, Germany. gallah@uni-muenster.de

used to interpret various cell-cell and cell-substrate interactions as well as different cell properties of cell monolayers.

11.2.1. Preparation of Solid Supported Membranes

Gold electrodes have been established as solid supports for impedance spectroscopy and quartz crystal microbalance experiments. These substrates are functionalized by different kinds of lipids or other substances containing a sulfhydryl group that chemically adsorbs to the gold surface, e.g., octanethiol. In a self-assembly process these substances build up a closest package of molecules on the surface of the gold electrode. In a second step phospholipids are applied as vesicle solution and spread on the surface giving a solid supported membrane. Proteins can be incorporated into these membranes by applying a mixture of proteins and lipid vesicles. Also, substrates functionalized with either a positive or a negative charge are used; oppositely charged lipids interact electrostatically to form the second layer of a bilayer membrane. Sulfhydryl functionalized lipids are directly used to immobilize membranes onto a gold support. For an overview of various self-assembly processes of lipid membranes onto solid supports see Ref. 3.

11.2.2. Physical Basics of Impedance Analysis

The electrical resistance of lipid membranes (double layers) is based on the hydrophobic core built by the acyl chains of lipids. This hydrophobic core is connected on both sides to the hydrophilic headgroups of the lipids, and therefore behaves electrically as a capacitor. Only few molecules are able to cross this barrier, e.g. water or oxygen. In contrast, larger or charged molecules are hindered of the transport. This barrier system is of exorbitant importance to maintain compartmentation within and between cells. The transport of non-diffusing molecules or ions across membranes is enabled by transport proteins. These can either act as carriers or membrane-spanning channels.

By using ions as substrates for transport proteins the ion flux can be characterized by a conductance due to the fact that ion transport corresponds to an electrical current. This current can be used to qualify and quantify the activity of transport proteins. Thus, a transport protein containing membrane is characterized by a resistance corresponding to the amount of transferred ions. In a simple model a lipid double layer can be mimicked by an equivalent electrical circuit (see Fig. 11.1A). The barrier properties are represented by the capacity C_m, whereas transport activities across the membrane are summarized as R_m. The parallel arrangement of C_m and R_m is related to the contemporary function of transport and barrier. The electrode/electrolyte interface is characterized by the formation of a Guy-Chapman layer with a high-capacitance C_{el} of about 33 $\mu F \cdot cm^{-2}$,[4] and the experimental setup of gold electrode, platinum electrode, and wires is summarized as resistor R_{el} (see Fig. 11.1A).

The complex electrical circuit with four elements CRRC is analyzed by an alternating current using various frequencies. A noninvasive voltage of several

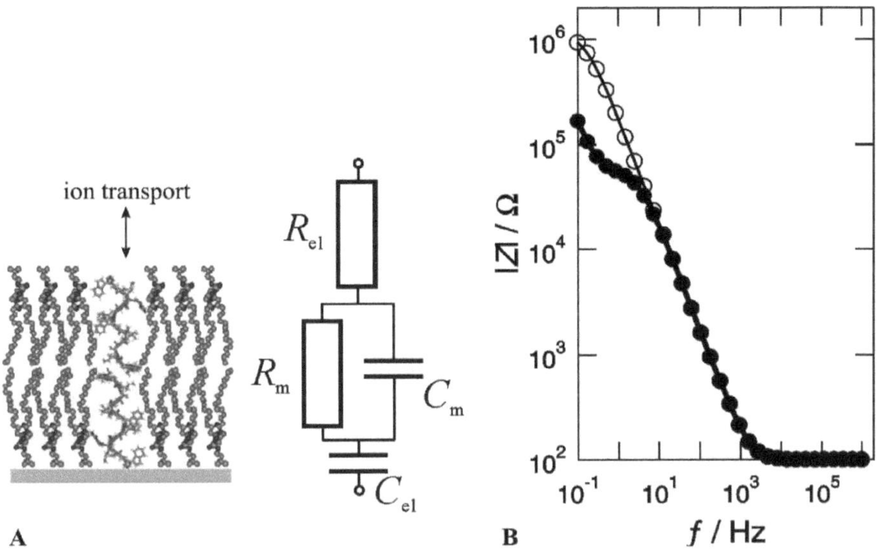

FIGURE 11.1. Impedance analysis of solid supported membranes. (A) Phospholipid membrane with an incorporated gramicidin channel and an equivalent electrical circuit. R_m and C_m represent the channel conductance and the barrier function of the membrane, respectively. (B) Impedance spectra of phospholipid membranes with (•) and without (○) gramicidin. The parameters are: $R_{el} = 100\ \Omega$, $C_{el} = 10\mu$ F, $C_m = 1\ \mu$ F, $R_m = 50$ kΩ (with gramicidin), and $R_m = 1$M Ω (without gramicidin) (for details, see Ref. 5).

millivolts is applied to the sample, and the complex impedance is measured as a function of the frequency (see Fig. 11.1B). A mathematical analysis using a transfer function for the total impedance $|Z|$ between voltage and current leads to the determination of the electrical parameters of lipid membranes.

$$|Z| = \sqrt{\left(R_{el} + \frac{R_m}{1 + (\omega C_m R_m)^2}\right)^2 + \left(\frac{1}{\omega C_{el}} + \frac{\omega C_m R_m^2}{1 + (\omega C_m R_m)^2}\right)^2} \qquad (11.1)$$

11.2.3. Impedance Spectroscopy of Solid Supported Membranes

Both electrical parameters of membranes, C_m and R_m, are analyzed to investigate different processes at the surface of membranes. The membrane capacity depends on the distance of the two polarized headgroup layers. Adsorption of protein mono- or multilayers lead to an increase of the distance between the juxtaposed polarized layers of the membrane, and as a result the membrane capacity decreases. The membrane resistance provides evidence for the transport of substances across membranes, and can therefore function as an analytical parameter to investigate for instance drug transport across model membranes. Impedance spectroscopy offers the possibility to analyze isolated transporters in a well-defined environment.

11.2.3.1. Protein Adsorption

The electrical parameters of solid supported membranes depend on, among other things, the membrane thickness. Therefore, the capacity and resistance of membranes vary with the adsorption of protein layers. Steinem et al. showed a concentration-dependent interaction of the membrane active bee venom melittin on solid supported membranes.[6] The protein adsorbs to the headgroups of the phospholipids resulting in a decrease of the membrane capacity and conductance. Both parameters grow with increasing protein concentration, which is evidence for a membrane penetration of the protein with a final lysis and the formation of mixed micelles at higher concentrations. In contrast, Salamon et al. demonstrated the membrane adsorption of penetratin and the incorporation of the protein into the headgroup region of the membrane, resulting in a decrease of membrane capacity without any change in membrane thickness.[7] Thus, the complexity of the mathematical transfer functions of impedance spectra facilitates the identification of single separated interactions of membrane active proteins.

11.2.3.2. Ion Transport

One of the main application of impedance spectroscopy is the characterization of transport properties of channel- or pore-forming proteins. The transport of charged substrates leads to a measurable current across the membrane. In contrast to black lipid membranes (BLM), solid supported membranes have a high long-term stability and an easy handling, which makes them suitable for easy ion transport assays. The smallest substrate, the proton, is exemplarily translocated by the light-sensitive bacteriorhodopsin.[8] The photocurrent of bacteriorhodopsin depends upon the protein concentration, the lipid environment, and pH. Increasing protein concentrations result in higher photocurrents. Also, negatively charged phospholipids have a positive effect on the photocurrent, whereas a maximum photocurrent exists at a pH of 6.4. No significant influence of the ionic strength was found on the translocation activity.[8] The family of gramicidins are alkaline transporters with dimeric conformation. Gramicidin D was analyzed by Ref. 5 in dioctadecyl-diamethyl-ammoniumbromide membranes on a gold electrode. As shown in Fig. 11.2, the membrane resistance decreases within the group of alkaline ions. This is caused by a hydration effect. The smaller the ion is, the stronger is its hydration sphere bound, leading to a decreasing radius of the hydrated ions within the group of alkaline ions. These two examples demonstrate the suitability of impedance spectroscopy to study ion transport across membranes and to characterize the involved proteins.

Solid supported membranes miss an aqueous compartment on one side of the membrane. This disadvantage has been overcome in a recent development given by Steinem and co-workers who used pore-spanning lipid membranes to enable

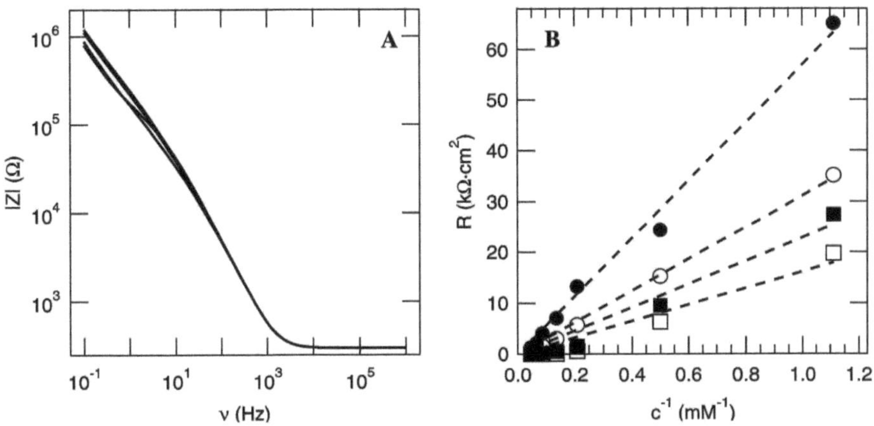

FIGURE 11.2. Impedance analysis of the alkaline ion transport through gramicidin D channels incorporated into dioctadecyl-dimethyl-ammoniumbromide membranes. (A) The total impedance decreases within the group of alkaline ions. (B) The analysis of the equivalent electrical circuit demonstrates the influence of the hydration radius of the ions on their conductivity. Within the group the hydration radius decreases leading to a higher conductivity within the gramicidin D channels. The symbols represent: • LiCl, ○ NaCl, ■ KCl, and □ CsCl. The parameters of the electrical equivalent are: $C_{el} = 9.4\ \mu F{\cdot}cm^{-2}$, $C_m = 1.1\ \mu F{\cdot}cm^{-2}$, $R_{el} = 100\ \Omega{\cdot}cm^2$, R_m and the Warburg impedance σ depend upon the alkaline ion (for details, see Ref. 5).

a liquid compartment on both sides of the membrane.[9] The authors developed a technique to immobilize solid supported membranes on porous alumina with pore diameters of about 20 to 50 nm. These "nano-black lipid membranes" (nano-BLM) have a long-term and mechanical stability, thus they are an improvement of common black lipid membranes, which have difficult preparation techniques and only a short-term stability. The successfull incorporation of channel proteins like gramicidin or alamethicin has been demonstrated.[10]

11.2.4. Impedance Spectroscopy of Cell Monolayers

The evaluation of the biological activity of pharmacological relevant substances is based on the application of cultured cells as sensor elements. Among suitable cell culture models competitive analytical methods are needed to detect cellular reactions as a function of pharmacological substances. The cell shape is a well-known criterion to analyze the cell fitness due to the fact that cells react very sensitive on changes of their environment. The method of electrical cell-substrate impedance sensing (ECIS) is a new electrochemical application to analyze changes in the cell shape as a function of biological, chemical, or physical stimuli without invasion of the cells.[11,12] Cell-cell and cell-substrate interactions can be characterized independently. Thus, the change of the cell shape is interpreted by electrical parameters.

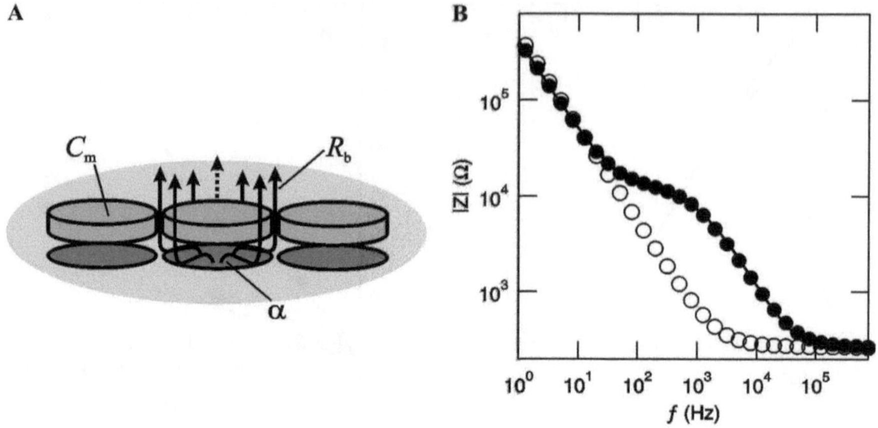

FIGURE 11.3. Electrical cell–substrate impedance sensing. (A) Scheme of the different impedance elements in cultured adherent cells. Cm represents the ion flux across the cell walls of the monolayer, α is the resistance between the cell monolayer and the substrate, and R_b is the resistance of the cell-cell contacts (for details see Ref. 13). (B) Impedance spectra of a gold electrode without (○) and with (•) a cell monolayer (for details, see Ref. 14).

11.2.4.1. Electrical Cell-Substrate Impedance Sensing

Confluent cell monolayers are cultured on gold electrodes, and the impedance be-tween a working electrode (several square millimeters) and a counter electrode (1000 times larger) is analyzed mathematically. There are three different ways for the electrical current from the working to the counter electrode (Fig. 11.3A).

(1) When leaving the working electrode the current has to pass the cleft between the lower side of cells and the surface of the electrode, which is characterized by the electrical parameter α. This cleft is often smaller than 0.1 μm, which makes it to the main part of the total impedance. Due to cellular adhesion and movement processes cells alter the distance between their lower side and the substrate, resulting in a change of α.

(2) Cell-cell contacts are also a kind of eye of a needle for the current, because the cleft between adjacent cells is also in the range of 0.1 μm. By stimulating cell-cell contacts the cleft can be opened or closed resulting in a change of the electrical parameter $_b$.

(3) Using alternating current with a high frequency (several kilo- or megahertz) the cells can be charged as a kind of capacitor. The capacity of the cell membrane C_m is a function of the cell shape, and changes of it are detected very sensitively. By applying a broad range of frequencies of the alternating current all three parameters, and corresponding changes in cell-cell and cell-substrate interactions or in the cell shape can be determined in a single experiment (Fig. 11.3B). Due to an electronic automatization and the parallele arrangement of eight electrodes, a set of experiments is possible as a function of time, simultaneously.

The expression of electrically tight intercellular junctions was proven by Arndt et al. who developed an apoptosis assay for endothelial cells from cerebral microvessels.[14] The authors used cycloheximide to cause apoptosis, which leads to a disassembly of barrier-forming tight junctions. The changes in cell-substrate contacts was then measured as a decrease in total impedance resulting from a decrease of both, α and R_b. Keese et al. developed a wound-healing assay for confluent cell monolayers after current treatment.[15] Cell monolayers on ECIS electrodes were subjected to currents resulting in severe electroporation and subsequent cell death. The authors were able to monitor cell migration into and ultimate healing of the wound as an increase in resistance and a decrease in capacitance of the cell monolayer.

11.3. Quartz Crystal Microbalance Analysis of Biomolecules

In 1959, Sauerbrey described the proportionality between the resonance frequency of quartz oscillators and their mass.[16] This was the key step for an analytical tool called "microbalance," bcause it was possible to analyze the thickness of thin solid films in a vacuum or air by means of quartz oscillators.[17–19] Unfortunately, the damping of water was too high to use quartz resonators as a bioanalytical tool.

In 1982, Nomura and Okuhara developed an oscillator circuit to compensate for this effect.[20] Measurements of, for example, protein adsorption were now possible. A recent development is the so-called "QCM-D" technique. Rodahl and co-workers established a technique to measure the resonance frequency of quartz crystals and the energy dissipation caused by water damping simultaneously.[21] The quartz resonators are stimulated by a frequency generator to oscillate with their resonance frequency, the source is then switched off resulting in a free oscillation of the resonators, and the damping of the resonators is detected. The resonance frequency is calculated and the complete procedure is continuously repeated every second. Basically shear wave resonators are used to analyze the binding of biomolecules to functionalized surfaces.

11.3.1. Basics of Piezoelectric Sensors

The application of an alternating electrical field to quartz crystals leads to a mechanical deformation of these crystals at their surface. This effect is called "inverse piezoelectricity." Periodical shifting of the surface charge of the resonators leads to an acoustic standing wave within the resonator perpendicular to its surface (see Fig. 11.4A). These quartz resonators are called "thickness shear mode" resonators (TSM). They are cut in AT orientation with an angle of 35.10° to the optical axis of a rod-shaped quartz crystal.

There is a strong correlation between mechanical or acoustical and electrical parameters in piezoelectric elements with the latter ones enabling direct determination. The correlation is given by the so-called "Butterworth-van-Dyke" analysis (BVD), which is described by Mason for unloaded resonators at an oscillation near to their resonance with only less damping.[22] Discrete impedance elements within the BVD circuit are interpreted as their mechanical analoga energy dissipation (R_q), mass loading (L_q) and elasticity (C_q) (see Fig. 11.4B and Ref. 23).

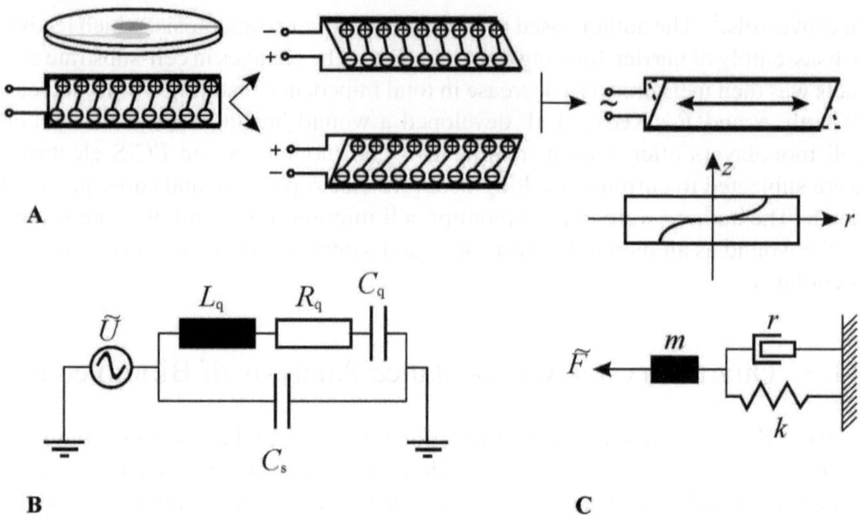

FIGURE 11.4. Equivalence of the mechanical and electrical properties of quartz resonators in quartz crsytal microbalance applications. (A) A periodic shifting of the surface charge of quartz crystals leads to a mechanical deformation, which appears as an oscillation of the resonators. An acoustic standing wave in the range of 10^6–10^7 Hz is generated perpendicular to the surface of the resonators. (B) The electrical analog is analyzed by a Butterworth-van-Dyke circuit. The inductivity L_q corresponds to the loaded mass, the resistor R_q to the damping factor, and the capacitor C_q represents the elasticity of the quartz resonator. The static capacity C_0 determines the admittance if the resonator oscillates far from its resonance frequency. (C) Mechanical scheme of an oscillating quartz with the driving force F, the loaded mass m, the damping factor r and the spring constant k of the oscillation.

The mass-sensing properties of quartz resonators are based on Sauerbrey's equation:

$$\Delta f = -\frac{2f_0^2}{A\sqrt{\bar{c}_{66}\rho_q}}\Delta m = -S_f \Delta m \qquad (11.2)$$

with the fundamental frequency f_0, the surface area A, the shear module $\bar{c}_{66}$, the density ρ_q, and the mass loading Δm of the resonator. Therefore, the change in resonance frequency of quartz oscillators is proportional to the loaded mass, with the integral mass sensitivity S_f composed only of quartz material constants. Under the assumption that the density of thin solid films and the quartz material is nearly identical, the adsorption of thin films of biomolecules leads to a decrease in resonance frequency.

Viscoelastic effects on the other hand have a strong influence on the oscillation of quartz resonators. A cell monolayer cultured on the gold electrode of a quartz resonator has different viscoelastic properties compared to pure water. Water behaves as a Newtonian fluid, but the flexible shape of cells and cell-cell as well as cell–substrate interactions lead to a viscoelastic inhomogeneous surface.

The investigation of both effects, mass loading and the change of viscoelastic properties, are demonstrated in different experiments.

11.3.2. Mass Loading onto Functionalized Surfaces

The interaction of different kinds of molecules can be characterized as a mass loading onto a functionalized surface. Some of these interactions are summarized in the next sections.

11.3.2.1. Adsorption of Proteins to Functionalized Surfaces

The adsorption of proteins to functionalized surfaces has been analyzed since the last decade starting with an investigation about the binding of human serum albumine (HSA) to quartz resonators.[24] The calculated integral mass sensitivity was significantly higher than the expected one from the Sauerbrey equation, and it was protein-dependent. This confusion was solved by Kasemo and co-workers who observed an energy dissipation during protein adsorption on functionalized surfaces investigating different forms of hemoglobin.[25–27] A two-step adsorption process is discussed with conformational changes within the protein monolayer and the bound water layers.

11.3.2.2. In Situ Hybridization of DNA/RNA on Quartz Resonators

One of the first bioanalytical applications using the quartz crystal microbalance technique was high-throughput investigations on genetic material. Quartz resonators were functionalized with single-stranded DNA bound to 11-sulfanyl-undecanol.[28] The complementary nucleotide strand hybridized with the immobilized one resulting in a decrease of resonance frequency. A common application is the modification of the phos-phate group at the 5′-end of the nucleotide to incorporate a sulfuryl group, which binds directly to the gold electrode on the surface of the quartz resonators.[29] Several techniques to immobilize funtionalized oligonucleotides are in practice, two of them are the use of peptide-nucleotides with a cysteine at the end of the oligonucleotide[30] similar to the sulfanyl group introduced by Su. A second application, especially for DNA or RNA, is the immbolization of negatively charged oligonucleotides by electrostatic interactions with a positively charged amine monolayer.[31] Janshoff et al. give an overview of DNA/RNA-hybridization techniques on quartz resonators.[17]

11.3.2.3. Lipid-Protein and Ligand-Receptor Interactions

A further development in the investigation of protein interactions at surfaces was given by the use of solid supported lipid membranes. Specific protein-receptor interactions are analyzed, and also protein adsorption to lipid model membranes. The bee venom melittin adsorbs in mono- and multilayers to phospholipid membranes, and the binding constant can be determined from a Langmuir adsorption isotherm to be between 20 and 100 μM, depending on the preparation technique of

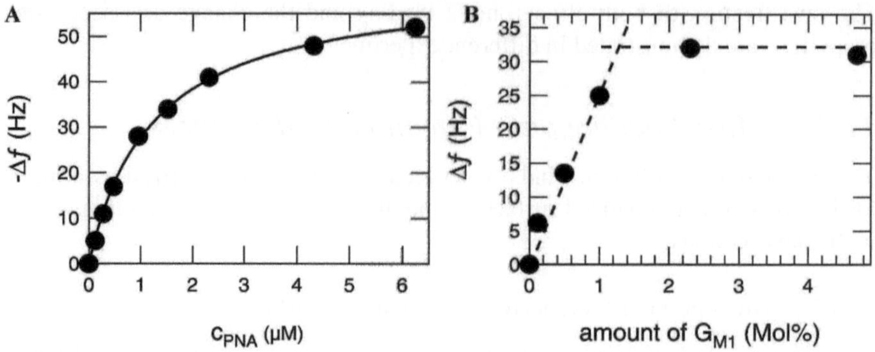

FIGURE 11.5. Binding of peanut agglutinin (PNA) to 1,2-palmitoyl-oleyl-sn-glycero-phosphatidylcholine (POPC) membranes containing differently functionalized gangliosides. (A) Langmuir adsorption isotherm of peanut agglutinin and POPC membranes containing 4.8 Mol% G_{M1}. (B) Langmuir adsorption isotherm of peanut agglutinin and POPC membranes containing 4.8 Mol% *asialo*-G_{M1}. The binding constants are $(0.83 \pm 0.04)\cdot 10^6$ M^{-1} for G_{M1} and $(6.5 \pm 0.3)\cdot 10^6$ M^{-1} for *asialo*-G_{M1} (for details, see Ref. 32).

the solid supported membranes.[33,34] In an investigation upon the heterotetrameric annexin A2t Ross et al. found a strong influence of the calcium ion concentration on A2t binding to phosphatidylserine containing membranes.[35] Typical time courses of annexin A2t binding to solid supported membranes are shown in Fig. 11.5A, the determination of the rate constants is demonstrated in Fig. 11.5B. While neglecting diffusion-limiting steps the adsorption of annexin A2t to solid supported membranes can be described as:

$$\Delta f\,(t) = \Delta f_{eq}(1 - e^{-k_s t}) \tag{11.3}$$

with the change in resonance frequence at equilibrium Δf_{eq}, and the protein concentration-dependent rate constant k_s, defined as

$$k_s = k^+ \cdot c + k^- \tag{11.4}$$

k^+ is the rate constant of adsorption, k^- the rate constant of desorption, and c the protein concentration. Both, kinetic and thermodynamic parameters of protein adsorption to solid supported membranes can be determined. The Ca^{2+}-dependent binding of annexin A1 to phosphatidylserine containing membranes was analyzed by Ref. 36. The authors were able to determine rate and affinity constants as a function of the calcium ion concentration in solution and the cholesterol concentration in the immobilized membranes. The decrease in calcium ion concentration leads to an increase of the rate of membrane binding of annexin A1, whereas cholesterol alters the affinity of annexin A1 to a high degree.

The determination of thermodynamic and kinetic parameters of ligand-receptor interactions was demonstrated by investigations of peanut agglutinin and ganglioside containing phospholipid membranes.[4,32,37] Binding constants of differently functionalized gangliosides like G_{M1} and *asialo*-G_{M1} revealed information about

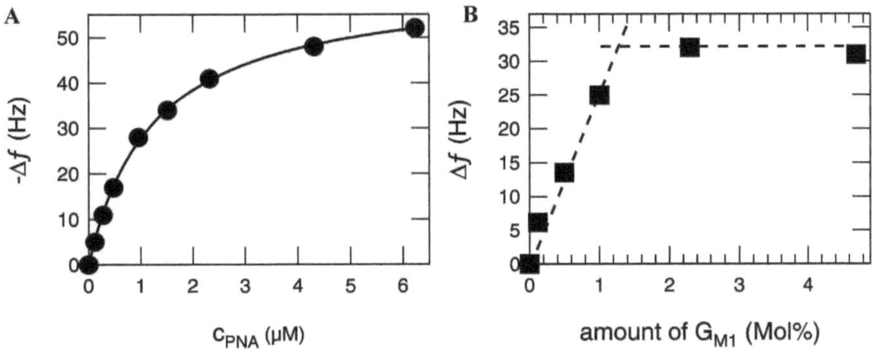

FIGURE 11.6. Binding of peanut agglutinin (PNA) to 1,2-palmitoyl-oleyl-sn-glycero-phosphatidylcholine (POPC) membranes containing differently functionalized gangliosides. (A) Langmuir adsorption isotherm of peanut agglutinin and POPC membranes containing 4.8 Mol% G_{M1}. The binding constant is $(0.83 \pm 0.04) \cdot 10^6$ M^{-1}. (B) Total frequency shift as a function of the receptor concentration. Equal PNA amounts were added and the saturation of the ligand-receptor interaction is at a concentration of about 1.3 Mol% of G_{M1} within POPC membranes (for details, see Ref. 52).

the influence of single substitutions at the receptor on its ligand binding properties (Fig. 11.6). N-acetylneuram acid acts as an inhibitor for peanut agglutinin binding, thus the binding constant of *asialo*-G_{M1} (lacking N-acetylneuram acid) is higher than that of G_{M1} (6.5 and $0.83 \cdot 10^6$ M^{-1}, respectively). The quantification of the inhibition of peanut agglutinin binding can solve the carbohydrate structure of the ganglioside receptor.[32] Because there is no unspecific binding of proteins to the immobilized lipid matrix on quartz resonators, and due to the regeneration of the well-defined surface functionalization, solid supported membranes on quartz resonators can be used as bioanalytical tools.

11.3.2.4. Immunosensors

Special ligand-receptor interactions, e.g., antigen-antibody, bacterial, or viral toxins and their receptors, can also be analyzed by so-called immunosensors. The surface of these sensors has to be functionalized adequately, the detection must be highly sensitive, specific and reproducible, and the surface of the sensors should be able to be regenerated.

König and Grätzel used immobilized antibody layers to detect the herpes virus in biological samples.[38] African swine fever virus and antibodies against it were analyzed by several groups. Abad et al.[39] reported an assay to detect antibodies against the protein p12 of African swine fever virus, whereas Uttenthaler and coworkers developed a sensor to immobilize antibodies on quartz resonators to detect virus particles.[40]

Also, several bacterial strains have been investigated by means of QCM, with an immobilized antibody layer as biosensor. *Vibrio cholerae* was analyzed by Carter et al., who used 10 MHz resonators.[41] Park and Kim used a thiolated antibody against *Salmonella typhimurium* to detect these bacteria,[42] and by investigating an antigen-antibody reaction Minunni et al. developed an assay to detect *Listeria monocytogenes*.[43] All assays are based on the antigen-antibody affinity of viruses or bacteria. This affinity is not limited to small organisms, but also organic molecules can be recognized by antibodies. Steegborn and Sklădal showed the determination of herbicide concentrations in a flow cell,[44] Liu et al. developed an assay to determine polycyclic aromatic hydrocarbons in solution.[45] All these examples show the broad variety of immunosensors based on the quartz crystal microbalance technique.

11.3.2.5. Enzymatic Activity at Functionalized Surfaces

In a recent study the synthase activity of a membrane-bound bacterial enzyme has been visualized by means of quartz crystal microbalance. Wältermann et al. showed that the formation of either wax esters or triacylglycerides at a lipid membrane is catalyzed by an enzyme called wax ester synthase/diacyl-glycerol-acyl-transferase (WS/DGAT).[46] The formation of lipid droplets, their maturation as well as their desorption from the lipid membrane were shown in a quartz crystal microbalance study (Fig. 11.7).

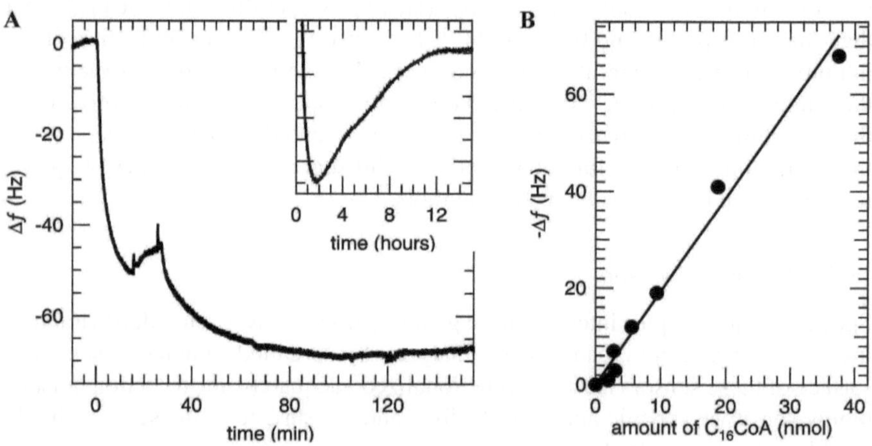

FIGURE 11.7. Enzymatic activity of a bacterial wax ester/diacyl-glycerol-acyl-transferse (WS/DGAT). Lipid droplet formation onto negatively charged phospholipid membranes is visualized by means of quartz resonators. (A) WS/DGAT is injected at $t = 0$ min. After 15 min the excess protein is removed. The two substrates, BSA stabilized hexadecanol and co-enzyme A activated palmitoyl, were added at $t = 30$ min. Transferase activity is observed for additional 90 min. Inset: After several hours, lipid droplets desorb from the lipid membrane. The enzymatic activity starts at $t = 0$ min when the substrates are injected. (B) The overall enzymatic activity can be analyzed as a function of the amount of co-enzyme A activated palmitoyl.[46]

11.3.3. Viscoelastic Effects of Cell Monolayers

The Sauerbrey equation facilitates the quantification of adsorbants onto functionalized quartz resonators. The proportionality of resonance frequency and loaded mass is in this case based on the effect of a rigid mass, namely, the water bar above the electrode. Its influence is nearly unaffected by the adsorption of biomolecules. This situation changes when investigating cells with the QCM technique. As already mentioned, viscoelastic effects cause a change in resonance frequency of quartz crystals. Under shear stress, cells on a quartz resonator do not behave homogeneously in their oscillation. In contrast, they oscillate like a viscoelastic body. This effect is used to analyze cell attachment and spreading onto functionalized surfaces mimicking parts of the extracellular matrix (ECM).

Three application modes have been developed to analyze cultured cells on a gold support. The QCM can be used in its normal (active) mode, driven by an oscillator circuit that determines continuously the resonance frequency. In the so-called passive mode, impedance spectra are measured near the resonance frequency (4.95–5.05 MHz). Due to its electromechanical coupling the resonance frequency of quartz crystals can be calculated using mathematical transfer functions. A combination of the passive impedance analysis of the quartz resonators and an impedance analysis of the cell monolayer facilitates the investigation of cell-cell and cell-substrate contacts simultaneously. The applications for all three running modes of quartz resonators and cell monolayers are described in the next sections.

11.3.3.1. Cell Attachment and Spreading

Cell adhesion was monitored as a function of seeding density (Fig. 11.8A). Cells of strains I and II of epithelial cell line MDCK, and Swiss-3T3 fibroblasts were

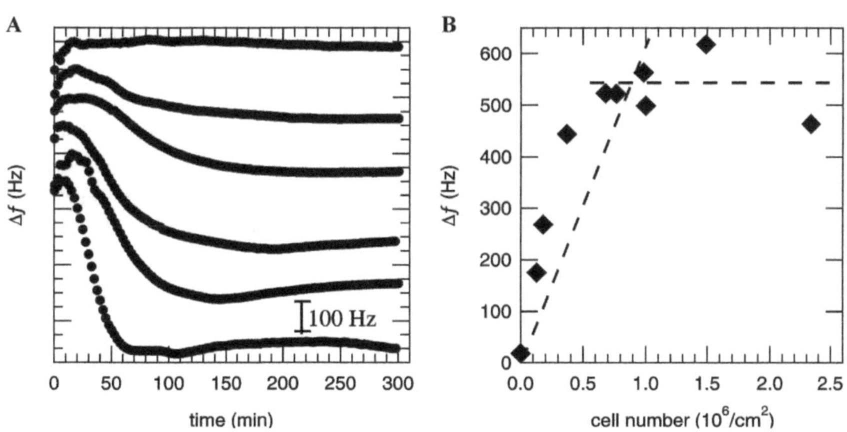

FIGURE 11.8. Monitoring cell adhesion with the quartz crystal microbalance. (A) Total shift of the resonance frequency as a function of the seeding density of MDCK-II cells. (B) Cell monolayer analysis. The horizontal line shows the maximum frequency shift corresponding to the formation of a confluent cell monolayer (for details, see Ref. 47).

allowed to settle and attach to resonators, and the adhesion characteristics were studied.[47] Upon formation of a confluent monolayer the frequency shift is proportional to the number of cells (positive slope in Fig. 11.8B), additional cells do not cause any frequency shift (horizontal line in Fig. 11.8B). The number of cells corresponds to the number determined by fluorescence micrographs. This sensitivity of the QCM for a cell monolayer was confirmed by Ref. 48 for osteoblasts.

Contacts between mammalian cells and artificial surfaces are mediated through focal contacts. These focal contacts are impaired by water-soluble peptides containing the amino acid sequence RGDS. Wegener et al. showed that derivatives of this short peptide inhibit cell adhesion to gold surfaces.[47] These experiments demonstrate the possibilities to study cell-ECM interactions by means of the quartz crystal microbalance.

11.3.3.2. Passive Mode

As already mentioned, the QCM can be operated in a passive mode by measuring impedance spectra near the resonance of quartz resonators. BVD equivalent circuits are analyzed and the electrical parameters are interpreted as their mechanical analoga. The damping of the resonators is characterized by the motional resistance R, and the mass loading is contributed to the inductance L. The change of both parameters is characteristic for each cell monolayer. Therefore, the adhesion kinetics and the influence of ECM com-ponents on cell attachment can be expressed in terms of ΔR and ΔL.[49] Interactions of the protein layer between the cell monolayer and the resonator contribute drastically to the oscillation of the resonators.[50]

11.3.3.3. Double Mode

By means of the QCM technique in the passive mode cell-cell contacts and cell morphology modulations are not observable. To overcome this limitation Wegener et al. developed an experimental setup to measure impedance spectra of the oscillating quartz and of the cell monolayer simultaneously.[51,52] The combination of impedance spectroscopy and electrical QCM is a powerful tool to monitor time-dependent changes in cell morphology (cell-cell interactions) as well as changes in cell-substrate contacts. For detailed information about the QCM as a noninvasive tool to investigate cell-substrate interactions *in situ*, the reader is referred to Ref. 53.

11.4. Discussion

The investigations of ligand-receptor interactions and the use of cell culture models have become a very powerful bioanalytic tool, especially in the case of pharmacological applications. The noninvasive character of both techniques facilitates the determination of various influences on cultured cells; thus the transport of pharmacologic substrates or the binding of these substrates to their receptors can be characterized qualitatively and quantitatively. This comprehensive overview demonstrates several useful applications, but further development will obtain even more suitable model systems to gain a more detailed view of molecular interactions at receptors and at the cell surface.

References

1. Ende, D., and Mangold, K.-M., 1993, Impedanzspektroskopie, *ChiuZ.* **27**:134–140.
2. Steinem, C., Janshoff, A., and Galla, H.-J., 2003, Biochemical applications of solid supported membranes on gold surfaces: Quartz crystal microbalance and impedance analysis, in: *Planar Lipid Bilayers (BLMs) and Their Application*, edited by H. T. Tien and A. Ottova, Elsevier, Amsterdam, 2003, pp. 991–1016.
3. Steinem, C., Janshoff, A., Ulrich, W.-P., Sieber, M., and Galla, H.-J., 1996, Impedance analysis of supported lipid bilayer membranes: a scrutiny of different preparation techniques, *Biochim. Biophys. Acta* **1279**:169–180.
4. Steinem, C., Janshoff, A., Wegener, J., Ulrich, W.-P., Willenbrink, W., Sieber, M., and Galla, H.-J., 1997c, Impedance and shear wave resonance analysis of ligand–receptor interactions at functionalized surfaces and of cell monolayers, *Biosens. Bioelectron.* **12**:787–808.
5. Steinem, C., Janshoff, A., Galla, H.-J., and Sieber, M., 1997a, Impedance analysis of ion transport through gramicidin channels incoporated in solid supported lipid bilayers, *Bioelectrochem. Bioenerg.* **42**:213–220.
6. Steinem, C., Galla, H.-J., and Janshoff, A., 2000, Interaction of melittin with solid supported membranes, *Phys. Chem. Chem. Phys.* **2**:4580–4585.
7. Salamon, Z., Lindblom, G., and Tollin, G., 2003, Plasmon-waveguide resonance and impedance spectroscopy studies of the interaction between penetratin and supported lipid bilayer membranes, *Biophys. J.* **84**:1796–1807.
8. Steinem, C., Janshoff, A., Höhn, F., Sieber, M., and Galla, H.-J., 1997b, Proton translocation across bacteri-orhodopsin containing solid supported lipid bilayers, *Chem. Phys. Lipids* **89**:141–152.
9. Drexler, J., and Steinem, C., 2003, Pore-suspending lipid bilayers on porous alumina investigated by electrical impedance spectroscopy, *J. Phys. Chem. B* **107**:11245–11254.
10. Römer, W., and Steinem, C., 2004, Impedance analysis and single-channel recordings on nano-black lipid membranes based on porous alumina, *Biophys. J.* **86**:955–965.
11. Giaever, I., and Keese, C.R., 1991, Micromotion of mammalian cells measured electrically, *Proc. Natl. Acad. Sci. USA* **88**:7896–7900.
12. Giaever, I., and Keese, C.R., 1993, A morphological biosensor for mammalian cells, *Nature* **366**:591–592.
13. Wegener, J., Hakvoort, A., and Galla, H.-J., 2000b, Barrier function of porcine choroid plexus elithelial cells is modulated by cAMP-dependent pathways *in vitro*, *Brain Res.* **853**:115–124.
14. Arndt, S., Seebach, J., Psathaki, K., Galla, H.-J., and Wegener, J., 2004, Bioelectrical impedance assay to monitor changes in cell shape during apoptosis, *Biosens. Bioelectron.* **19**:583–594.
15. Keese, C.R., Wegener, J., Walker, S.R., and Giaever, I., 2004, Electrical wound-healing assay for cells *in vitro*, *Proc. Natl. Acad. Sci. USA* **101**:1554–1559.
16. Sauerbrey, G., 1959, Verwendung von Schwingquarzen zur Wägung dünner Schichten und zur Mikrowägung, *Z. Phys.* **155**:206–222.

17. Janshoff, A., Galla, H.-J., and Steinem, C., 2000, Piezoelectric mass-sensing devices as biosensors—an alterna-tive to optical biosensors? *Angew. Chem. Int. Ed.* **39**:4004–4032.
18. Schumacher, R., 1990, The quartz crystal microbalance: a novel approach to the in-situ investigation of interfacial phenomena at the solid/liquid junction, *Angew. Chem. Int. Ed. Engl.* **29**:329–343.
19. Buttry, D.A., and Ward, M.D., 1992, Measurement of interfacial processes at electrode surfaces with the elec-trochemical quartz crystal microbalance, *Chem. Rev.* **92**:1355–1379.
20. Nomura, T., and Okuhara, M., 1982, Frequency shifts of piezoelectric quartz crystals immersed in organic liquids, *Anal. Chim. Acta* **142**:281–284.
21. Rodahl, M., Höök, F., Krpzer, A., Brzezinski, P., and Kasemo, B., 1995, Quartz crystal microbalance setup for frequency and Q-factor measurements in gaseous and liquid environments, *Rev. Sci. Instrum.* **66**:3924–3930.
22. Mason, W.P., 1965, Physical Acoustics, Vol. 2A, Academic Press, New York.
23. Rosenbaum, J.F., 1988, *Bulk Acoustic Wave Theory and Devices*, Artechhouse, Boston.
24. Muratsugu, M., Ohta, F., Miya, Y., Hosokawa, T., Kurosawa, S., Kamo, N., and Ikeda, H., 1993, Quartz crystal microbalance for the detection of microgram quantities of human serum albumin: relationship between the frequency change and the mass of protein adsorbed, *Anal. Chem.* **65**:2933–2937.
25. Höök, F., Rodahl, M., Brzezinski, P., and Kasemo, B., 1998a, Energy dissipation kinetics for protein and anti-body-antigen adsorption under shear oscillation on a quartz crystal microbalance, *Langmuir* **14**:729–734.
26. Höök, F., Rodahl, M., Kasemo, B., and Brzezinski, P., 1998b, Structural changes in hemoglobin during adsorp-tion to solid surfaces: Effects of pH, ionic strength, and ligand binding, *Proc. Natl. Acad. Sci. USA* **95**:12271–12276.
27. Höök, F., Rodahl, M., Brzezinski, P., and Kasemo, B., 1998c, Measurements using the quartz crystal micro-balance technique of ferritin monolayers on methyl-thiolated gold: dependence of energy dissipation and saturation coverage on salt concentration, *J. Colloid Interface Sci.* **208**:63–67.
28. Su, H.A., 1991, Dissertation, University of Toronto.
29. Okahata, Y., Matsunobu, Y., Ijiro, K., Mukae, M., Murakami, A., and Makino, K., 1992, Hybridization of nucleic acids immobilized on a quartz crystal microbalance, *J. Am. Chem. Soc.* **114**:8299–8300.
30. Wang, J., Nielsen, P.E., Jiang, M., Cai, X., Fernandes, J.R., Grant, D.H., Ozsoz, M., Beglieter, A., and Mowat, M., 1997, Mismatch-sensitive hybridization detection by peptide nucleic acids immobilized on a quartz crystal microbalance, *Anal. Chem.* **69**:5200–5202.
31. Caruso, F., Rodda, E., Furlong, D.N., Niikura, K., and Okahata, Y., 1997, Quartz crystal microbalance study of DNA immobilization and hybridization for nucleic acid sensor development, *Anal. Chem.* **69**:2043–2049.
32. Janshoff, A., Steinem, C., Sieber, M., and Galla, H.-J., 1996a, Specific binding of peanut agglutinin to G_{M1}-doped solid supported lipid bilayers investigated by shear wave resonator measurements, *Eur. Biophys. J.* **25**:105–113.
33. Okahata, Y., Ebara, Y., and Sato, T., 1995, The quartz-crystal-microbalance study of protein binding on lipid monolayers at the air-water interface, *MRS Bull.* **20**:52–56.
34. Steinem, C., Janshoff, A., and Galla, H.-J., 1998, Evidence for multilayer formation of melittin on solid-supported phospholipid membranes by shear-wave resonator measurements, *Chem. Phys. Lipids* **95**:95–104.

35. Ross, M., Gerke, V., and Steinem, C., 2003, Membrane composition affects the reversibility of annexin A2t binding to solid-supported membranes: a QCM study, *Biochemistry* **42**:3131–3141.
36. Kastl, K., Ross, M., Gerke, V., and Steinem, C., 2002, Kinetics and thermodynamics of annexin A1 binding to solid-supported membranes: a QCM study, *Biochemistry* **41**:10087–10094.
37. Janshoff, A., Steinem, C., Sieber, M., el Bayâ, A., Schmidt, M.A., and Galla, H.-J., 1997, Quartz crystal micro-balance investigation of the interaction of bacterial toxins with ganglioside containing solid supported membranes, *Eur. Biophys. J.* **26**:261–270.
38. König, B., and Grätzel, M., 1994, A novel immunosensor for Herpes viruses, *Anal. Chem.* **66**:341–344.
39. Abad, J.M., Pariente, F., Hernández, L., and Lorenzo, E., 1998, A quartz crystal microbalance assay for detec-tion of antibodies against the recombinant African swine fever virus attachment protein p12 in swine serum, *Anal. Chim. Acta* **368**:183–189.
40. Uttenthaler, E., Kößlinger, C., and Drost, S., 1998, Characterization of immobilization methods for African swine fever virus protein and antibodies with a piezoelectric immunosensor, *Biosens. Bioelectron.* **13**:1279–1286.
41. Carter, R.M., Mekalanos, J.J., Jacobs, M.B., Lubrano, G.J., and Guilbault, G.G., 1995, Quartz crystal microbal-ance detection of *Vibrio cholerae* O139 serotype, *J. Immunol. Methods* **187**:121–125.
42. Park, I.-S., and Kim, N., 1998, Thiolated Salmonella antibody immobilization onto the gold surface of piezo-electric quartz crystal, *Biosens. Bioelectron.* **13**:1091–1097.
43. Minunni, M., Mascini, M, Carter, R.M., Jacobs, M.B., Lubrano, G.J., and Guilbault, G.G., 1996, A quartz crystal microbalance displacement assay for *Listeria monocytogenes*, *Anal. Chim. Acta* **325**:169–174.
44. Steegborn, C., and Skládal, P., 1997, Construction and characterization of the direct piezoelectric immunosensor for atrazine operating in solution, *Biosens. Bioelectron.* **12**:19–27.
45. Liu, M., Li, Q.X., and Rechnitz, G.A., 1999, Flow injection immunosensing of polycyclic aromatic hydrocar-bons with a quartz crystal microbalance, *Anal. Chim. Acta* **387**:29–38.
46. Wältermann, M., Hinz, A., Robenek, H., Troyer, D., Reichelt, R., Malkus, U., Galla, H.-J., Kalscheuer, R., Stöveken, T., von Landenberg, P., and Steinbüchel, A., 2004, Mechanism of lipid-body formation in pro-karyotes: how bacteria fatten up, *Mol. Microbiol.* **55**: 75–763.
47. Wegener, J., Janshoff, A., and Galla, H.-J., 1998, Cell adhesion monitoring using a quartz crystal microbalance: comparative analysis of different mammalian cell lines, *Eur. Biophys. J.* **28**:26–37.
48. Redepenning, J., Schlesinger, T.K., Mechalke, E.J., Puleo, D.A., and Bizios, R., 1993, Osteoblast attachment monitored with a quartz crystal microbalance, *Anal. Chem.* **65**:3378–3381.
49. Reiss, B., Janshoff, A., Steinem, C., Seebach, J., and Wegener, J., 2003, Adhesion kinetics of functionalized vesicles and mammalian cells: a comparative study, *Langmuir* **19**:1816–1823.
50. Wegener, J., Seebach, J., Janshoff, A., and Galla, H.-J., 2000a, Analysis of the composite response of shear wave resonators to the attachment of mammalian cells, *Biophys. J.* **78**:2829–2833.

51. Wegener, J., Sieber, M., and Galla, H.-J., 1996, Impedance analysis of epithelial and endothelial cell monolayers cultured on gold surfaces, *J. Biochem. Biophys. Methods* **32**:151–170.
52. Janshoff, A., Wegener, J., Sieber, M., and Galla, H.-J., 1996b, Double mode impedance analysis of epithelial cell monolayers cultured on shear wave resonators, *Eur. Biophys. J.* **25**:93–103.
53. Wegener, J., Janshoff, A., and Steinem, C., 2001, The quartz crystal microbalance as a novel means to study cell-substrate interactions in situ, *Cell Biochem. Biophys.* **34**:121–151.

Index

MIX
Papier aus verantwortungsvollen Quellen
Paper from responsible sources
FSC® C105338

If you have any concerns about our products,
you can contact us on
ProductSafety@springernature.com

In case Publisher is established outside the EU,
the EU authorized representative is:
Springer Nature Customer Service Center GmbH
Europaplatz 3, 69115 Heidelberg, Germany

Printed by Libri Plureos GmbH
in Hamburg, Germany